The Complete Guide to

WRITING & PRODUCING
TECHNICAL MANUALS

The Complete Guide to

WRITING & PRODUCING TECHNICAL MANUALS

LESLIE M HAYDON

A Wiley-Interscience Publication

JOHN WILEY & SONS, INC

New York • **Chichester** • **Brisbane** • **Toronto** • **Singapore**

This text is printed on acid-free paper.

Library of Congress Cataloging-in-Publication Data:
Haydon, Leslie M.
    The complete guide to writing and producing technical manuals
/ Leslie M Haydon.
        p.   cm.
    "A Wiley-Interscience publication."
    Includes index.
    ISBN 0-471-12281-5 (paper : acid-free paper)
    1. Communication of technical information. 2. Technology—
Documentation. 3. Technical manuals. I. Title.
    T10.5.H39   1995
    808'.0666—dc20   94-49677

Printed in the United States of America

10 9 8 7 6 5 4 3 2

PREFACE

For more than 30 years, I have been regularly involved with technical manuals, first as a user, then in the role of a technical publications writer/editor. As a user, the technical manuals that I have had to rely on for my engineering information has ranged in quality from exceptional to extremely poor. These latter manuals, generally prepared on a shoestring, have had shortcomings in almost every facet of their presentation.

By contrast, professional and technical handbooks for the bigger industries like aerospace, electronics or mainframe computer systems, for example, are quite consistent in all aspects of their presentation. This is mainly due to their publications being edited and prepared by professional publications organizations or their own in-house technical publications departments.

Unfortunately, many technical manuals being produced today are prepared in-house by company staff and are often little more than an organized collection of data sheets, engineering reports and drawings with some explanatory text thrown in for good measure. In a great number of instances, the company cannot be unduly criticized because there are very few publications available that will provide detailed advice in the preparation, assembly and printing of their technical manual. The hapless personnel, tasked with the manual's preparation, may be very skilled in using desktop publishing techniques, but without proper guidance in manual preparation are generally forced to make it up as they go along!

The availability of technical manual preparation handbooks in our ever-increasing technological world appears to be inversely proportional to the number of complex technical equipments that are entering our lifestyle by the minute. Conversely, there is an overabundance of books detailing how to write and edit technical English for computer software, hardware and computerized operating systems.

To my knowledge there are very few reference books on technical manual preparation on the market today that will provide a complete guide to the many aspects of actually merging together the proper formats, text and illustrations needed to create a good technical manual.

This lack of good style manuals was brought home to me many years ago when I attempted to locate and recommend such a book to a less experienced colleague who was to succeed me as a technical writer/manuals co-ordinator in a large electronics manufacturing company. Having learnt my trade over many years of hands-on experience, I never really needed such a reference, although in hindsight I believe that it would have made life somewhat easier.

One thing that has always disturbed me is the atrocious documentation that accompanies many very expensive complex items of equipment being marketed today. With the advent of computer-based desktop publishing hardware and the multitude of software packages in this field, companies have started producing their own documentation, unfortunately without much experience or proper guidance. I cannot determine whether this is a make-do attitude on the part of the manufacturer to pare some costs from the production or simply an inability to know what is needed—it is likely a combination of both factors. There is no need for this to continue.

It became very apparent to me that there was a serious shortage of handbooks providing even the basic rudiments of technical manual preparation. This book is my effort to correct this oversight.

Leslie Haydon
1995

ACKNOWLEDGEMENTS

Producing a reference book on a subject of this magnitude cannot be the work solely of a single author, nor can it be an original work. I would like to express my heartfelt thanks to the many people and companies whose original works have been an inspiration to the creation of this handbook. It has been a challenging, stimulating and satisfying effort and through the course of its creation, even I have learnt many new things.

Grateful acknowledgement is made to the following companies, publishers, and agents for their permission to use and adapt copyrighted materials:

John Wiley & Sons of New York, N.Y. for permission to use material from the *Handbook of Technical Writing Practices*, edited by John Stello, 1971; Domtar Fine Papers Limited for use of material from *The Fundamentals of the Printing and Duplication Processes*; The AIP Publication Board, for kind permission to use material from the *AIP Style Manual, Fourth Edition 1990*; Pitman Publishing of London, England for the use of material from *The Preparation and Production of Technical Handbooks*, by H. Heys, 1965; General Electric Aircraft Engines for permission to use text from their publication, *Style Manual For Technical Writers*, 1964; General Motors Canada Limited and Hyundai Auto Canada Inc., for use of drawings from their vehicle owner's handbooks; and the many other companies who willingly gave permission to use their marketing material.

Finally, I would like to pay tribute to the many individuals who have assisted me in various ways; at Wiley-Interscience, my editor, George Telecki, for his guidance through the preparation stages and especially Christina Della Bartolomea for her meticulous final editing, resulting in a much improved text; Roger Freeman, for his constructive comments and enthusiasm for the book; my colleagues at Ontario Hydro, in particular Richard Qua and Kemp Reese for their helpful advice and the use of their fine drawings; Bill Barlow for his photographic help, and my wife, for her encouragement over the years I have labored on this project.

TRADEMARK ACKNOWLEDGEMENTS

Adobe Illustrator is a registered trademark of Adobe Systems, Inc.

Apple, Macintosh are registered trademarks of Apple Computer, Inc.

Apollo is a registered trademark of Apollo Computer, Inc., a subsidiary of Hewlett-Packard Co.

Compaq is the registered trademark of Compaq Computer Corporation.

Epson is a registered trademark of Epson America Inc.

Helvetica and *Times* are registered trademarks of Allied Linotype Corporation.

HP and *Laserjet* are registered trademarks of the Hewlett-Packard Company.

Interleaf is a registered trademark of Interleaf, Inc.

Pagemaker is a registered trademark of Aldus Corporation.

Quark Xpress is a registered trademark of Quark, Inc.

SUN is a registered trademark of Sun Microsystems, Inc.

Ventura Publisher is a registered trademark of Ventura Software, Inc.

WordPerfect is a registered trademark of the WordPerfect Corporation.

Microsoft Word is a registered trademark of Microsoft, Inc.

Zerox is a registered trademark of the Zerox Corporation.

CONTENTS

1—INTRODUCTION 1

2—TECHNICAL WRITING AS A CAREER 3
The Technical Writer Defined 3
The Art of Technical Writing 4
The Training of Technical Writers 4
Employment Opportunities 6

3—TECHNICAL MANUALS AND HANDBOOKS 7
Types of Manuals 8
General Manuals 8
Custom-Made Manuals 9
Arrangement of Technical Manuals 10
General Information 11
Operating Instructions 12
Theory of Operation 12
Maintenance Instructions 12
Spare Parts 16
Installation 16

4—PLANNING A TECHNICAL MANUAL 19
Initial Decisions 19
Definitions 19
Manual Structure 20
Manual Production Plan 26

5—PUBLISHING SYSTEMS 31

6—LAYOUT AND FORMAT 33
Introduction 33
Typefaces and Typestyles 33
Legibility 33
Readability 34
Typographic Terminology 34

Page Layout 38
Text Layout 39
Paragraphs 39
Headings 40
Numbering 44
  Paragraph Numbering 44
  Page Numbering 45
  Figure and Table Numbering 46
  Front Matter Numbering 46
  Blank Page Numbering 46
Warnings and Cautionary Notices 47
Titles and Logotypes 48

7—MANUAL WRITING STYLE 49
Introduction 49
Operation 50
  Front Matter 50
  Introduction 54
  Description 55
  Operation 55
  Troubleshooting 57
  Minor Servicing 58
Maintenance 58

8—PREPARING A MANUAL SPECIFICATION 71
Introduction 71

9—FRONT MATTER AND INTRODUCTORY MATERIAL 95
The Cover Page 96
Artificial Respiration 96
Electric Shock—Action and Rescue 96
Safety Instructions 96
Warranty 101
Table of Contents 101
List of Figures (Illustrations) 101
List of Tables 105
Amendment Record 105
List of Effective Pages 105

10—ILLUSTRATIONS 107
Introduction 107
Types of Illustrations 110
  Line Drawing 110
  Charts 110
  Graphs 115
  Captions and Labels 115
  Choosing Scales and Grids 119
Tips for Creating and Using Illustrations 119

11—TABLE PREPARATION 121
   Introduction 121
   Guidelines for Creating and Using Tables 121
      Informal Tables 123
      Tips for Creating Tables 125

12–OPERATION 127
   Introduction 127
   Content 127

13—MAINTENANCE AND REPAIR INSTRUCTIONS 131
   Introduction 131
   Equipment Configurations 132
   Safety Instructions 133
   Modes of Operation 134
   Theory of Operation 134
   Test Equipment and Tools 135
   Lowest Level of Service 135
   Inspection Schedules and Procedures 136
   Periodic and Preventative Maintenance 138
   Trouble Analysis 138
   Removal and Replacement of Parts 139
   Adjustment and Alignment 140
   Performance Checks 142
   Drawings 142

14—ILLUSTRATED PARTS BREAKDOWN 143
   Introduction 143
   Numerical Indexes 144
   Reference Designation Indexes 144
   Manufacturer's Codes 144
   Military IPB Preparation 147

15—APPENDIXES AND ADDENDA 149
   Introduction 149
   The Appendix 149
   The Addendum 150
   Integration 151
   Layout 151
   Referencing 151

16—AMENDING MANUALS 153
   The Need for Amendments 153
   Incorporating Amendments 153
   Distributing and Recording Amendments 154
   Amendment Identification 155
   Amendment Control 156

17—PREPARING CAMERA-READY COPY 157
   Introduction 157
   Text Page Preparation 157
   Preparing Illustrations 158

18—PRINTING AND BINDING 161
   Introduction 161
   Printing Papers 161
      Selection Criteria 161
      Paper Grades 162
      Finish 163
      Grain 163
      Weight 164
      Folding 164
      Types of Folds 165
      Collating 167
      Stitching 167
      Binding Methods 168
   Bookbinding 171
   Finishing 173
   Tabbed Index Dividers 174
   Shrink Packaging 174

19—THE TECHNICAL EDITOR 175
   Introduction 175
   The Role of the Technical Editor 177
   Style Manuals 179
   Editing Procedures 181
   Principles of Editing 181

20—A TECHNICAL HANDBOOK DEPARTMENT
   – From Concept to Operation 187
   Introduction 187
   Establishing the Department 188
   Layout of the Department 190
      Technical Writers 190
      Illustrators 190
      Allocation of Files 191
   Furniture 191
   Contact with Internal Groups 191
      Engineering Departments 192
      Illustrating and CAD/CAM Departments 192
      Service Department 192
      Sales, Publicity, and Training Departments 193
      Photographic Department 193
      Contract, Patents, and Specifications Department 193
   Contact with External Groups 193
      Client Relationships 194
      Printers and Suppliers 194
      Contract and Temporary Staff 195
   Time and Cost Factors 195

APPENDIXES

A   Capitalization Rules 197
B   Mathematical and Scientific Terminology 207
C   Using the Metric (SI) System 217
D   Numbers in Technical Manuals 227
E   Abbreviations 235
F   Footnotes 237
G   Punctuation 245

GLOSSARY OF TECHNICAL TERMS 265

BIBLIOGRAPHY 277

INDEX 279

LIST OF FIGURES

4-1   Basic Manual Structure   21
4-2   Manual Structure Chart   27
4-3   Production Plan – Block Diagram Style   28
4-4   Critical Path Monitoring   30
6-1   Dual Column Setup with Table and Graphic   40
6-2   Full Page, Narrow Column Layout   41
6-3   Full Page, Wide Column Layout   43
7-1   Typical Manual Title Page   52
7-2   Typical Warranty Statement   53
7-3   Typical Operating Controls   56
7-4   Block Diagram Development   63
7-5   Fault Diagnosis Chart   66
7-6   Troubleshooting Chart   68
9-1   Typical Cover Page   97
9-2   Artificial Respiration Page   98
9-3   Typical Safety Instruction Sheet   100
9-4   Typical Warranty Statement   102
9-5   Typical Table of Contents (First Page)   103
9-6   Typical Table of Contents (Last Page)   104
9-7   Revision Block   105
10-1   Assembly View   109
10-2   Operational Illustration   109
10-3   Animated Illustration   110
10-4   Cartoon-Style Illustration   110
10-5   Location View   111
10-6   Phantom View   111
10-7   Typical Lubrication Diagram and Chart   112

10-8   Typical Waveform Diagram 112
10-9   Typical Wiring Diagram 113
10-10 Typical Schematic Diagram 113
10-11 Typical Cutaway Diagram 114
10-12 Typical Hand-Drawn Line Drawing 116
10-13 Typical Pie Chart Illustration 117
10-14 Typical Bar Chart Illustration 117
10-15 Typical Pictorial Chart 118
11-1   Nomenclature for the Parts of a Table 123
11-2   Complex Table Form 125
12-1   Typical Operating Controls 130
13-1   Servicing Test Setup 136
13-2   Disassembly Illustration 141
14-1   Parts List Page 145
14-2   An Illustrated Parts Breakdown 146
14-3   A Numerical Index of Parts 147
14-4   Reference Designation Index 148
14-5   Manufacturer's Code List 148
15-1   Appendix Front Page 152
16-1   Revision Block With Typical Entries 155
17-1   Illustration Rework 159
18-1   Common Types of Paper Folds 166
18-2   Stitching Styles 167
18-3   D-Ring Binder with Pockets 169
18-4   Typical Loose-Leaf Binders 169
18-5   Spiral and Plastic Binding 170
18-6   Plastic Binding Machine 171
18-7   A Thermal Binding Machine 172
18-8   Perfect Binding 172
18-9   Hand-Sewn Binding 173

The Complete Guide to

WRITING & PRODUCING
TECHNICAL MANUALS

CHAPTER 1

INTRODUCTION

The intent of this reference handbook is to provide guidance in the techniques and procedures necessary to produce a high-quality technical manual. The cosmetic appearance of the finished product will be directly related to the facilities available for preparing the manual; however, the quality of the manual can only be judged on the technical material within its covers and the way it is presented to the reader.

Whether the technical manual is prepared on a sophisticated Linotronic typesetting system with the accompanying high resolution or a simple desktop publishing system using inexpensive page composition software such as *WordPerfect*® and a good laser printer, it is the technical content and the layout of it that are the most important aspects. Given the same technical content, a high-gloss brochure-like handbook with poorly written but essentially correct text is of lesser value than a simple photocopy-quality economically bound booklet with that information laid out in an informative, detailed technical format.

It has been usual to find, within small engineering companies at least, that the production of a technical manual has been the responsibility of an engineer or technician, neither of whom have had any experience in the field of publication preparation. Generally these people are adequately suited to provide a technical manuscript suitable for a manual; however, there is little material in print to guide them in the preparation of the manual in terms of format, paper types, printing procedures, and so on. Unless the draft manuscript is handed over to a professional technical publications house for preparation, or the company itself has an in-house publications department, the resultant manual often finishes up as a photocopied reproduction, poorly presented, disorganized, and more often than not, reflecting a poor image of the company itself.

This handbook guides the reader through a series of chapters, each devoted to a particular subject in the process of preparing a technical manual. It provides comprehensive coverage of such topics as manual planning, format preparation, specification development, preparation of camera-ready copy, diagrams, charts, tables, parts lists, word processing and typesetting equipment, printing and binding options, and many other aspects relevant to the preparation of a technical manuscript for publication.

Where possible, options are suggested so that the program can be tailored to suit equipment the company has or any company restrictions. Some services available from commercial photographic and reproduction companies are discussed to enable the reader to readily assess their capabilities and utilize their services more efficiently.

Very little of this book is devoted to the art of technical writing itself as this subject has been covered adequately in many fine works over the years; however, appendixes have been included, which cover a series of subjects pertaining to aspects of manual text presentation. These include capitalization, the presentation of mathematics and scientific terminology, the correct use of metric notation, numbers, abbreviations, and finally some aspects of punctuation.

The next chapter introduces the art, career, and training of the technical writer, without whom the technical publication would seldom make the necessary step from a collection of notes to a successful publication.

CHAPTER 2

TECHNICAL WRITING AS A CAREER

This volume would not be complete without some space being devoted to the role of the technical writer in the preparation of a technical manual. It is the skill of this specialist that provides the foundation on which a successful technical publication is created, alongside the needs and aspirations of the company to whom the technical writer dedicates his or her expertise.

It is worthwhile to look at some aspects of the technical writing profession and examine it as a potential career goal.

THE TECHNICAL WRITER DEFINED

Essentially the technical writer is a technically orientated individual with a high degree of skill in the field of written communications.

In his paper, *Education for Technical Writers*, Dr. John A. Walter, has depicted a rather idealistic portrayal of the technical writer (based on a survey of some 160 U.S. companies): "It is fairly easy to picture the ideal technical writer/editor. Armed with a thorough knowledge of acceptable current usage, backed up by a sound grasp of the rules of grammar, spelling and punctuation, he must be adept at taking the rough copy of the scientist or engineer and turning it into readable prose – readable, that is, for the intended reader. This skill must be buttressed by sound knowledge of the technical material dealt with so that he may not only be able to increase the readability of the paper he edits but also catch technical flaws the author inadvertently overlooked.

Throughout this activity he must remain cooperative, gracious and unruffled in dealing not merely with the author but also with his supervisor and those members of management who are interested in the project.

Moreover, he must show himself knowledgeable in correlating his work and that of the author with that of the illustrators assigned to the project. Throughout the entire activity, he will demonstrate that he is conversant with all of the problems of getting a completed manuscript through production so that the finished product will be one which author, publications supervisor, the printer, management officials, not to mention the customer, can be proud."

Obviously few technical writers can lay claim to meeting all of these requirements, but it does provide a benchmark with which to judge one's own performance or strive to attain.

THE ART OF TECHNICAL WRITING

So that we may better understand the role of the technical writer in the industrial environment, we should first examine the art of technical writing itself.

Technical writing is a form of written communication that conveys scientific and technical information in a clearly defined and accurate form. The level of writing and editing—and both are equally important in this field—can range in depth from simple uncomplicated narratives to technical papers with a high degree of technological complexity and terminology.

All aspects of science and technology must be recorded and the information disseminated in a manner conducive to the advancement of that field. It is the art of technical writing that bridges this communications gap.

THE TRAINING OF TECHNICAL WRITERS

Technical writers are found in a wide range of technological environments, for example, aeronautical, electronic, power, petrochemical, or medicine, so the subject matter dealt with is substantially diversified. The range of tasks within these fields is extensive: report writing, proposal and brochure preparation; technical manual editing and preparation, these are but a small cross-section of the areas with which the technical writer can be involved.

To meet current requirements, technical writers are expected to have, at the very least, a sound technical background in the discipline they are writing for, coupled with several years experience in formal technical writing and editing.

The obvious question that follows is: How are these requirements attained? If we use the field of electronic engineering as an example of the areas of specialization, it is soon obvious that it contains a range of subdisciplines of almost staggering proportions: analog, digital, computers, medical, industrial, radar, communications, and avionics, to name the more common ones. The technology of electronics is so complex that even practicing engineers themselves cannot hope to specialize in any more than two or three fields at one time.

While technical writers are not expected to be engineers, the requirement that their technical knowledge be related to the field of company endeavor, virtually eliminates the possibility of having formal technical training programs available for every interest. There are several paths available for those wishing to gain some level of formal training on which to base a career in technical writing:

a. University: As an undergraduate major in an appropriate science with a minor in English.

b. University: As an undergraduate major in English with minor in a science.

c. Institutes of technology: As a graduate technologist with electives in English, Report Writing, Journalism, and so on. (This latter subject would likely have to be taken as an extension course.)

While such fundamental training requirements for the technical writer can be specified with some degree of confidence, the controversy of whether the technical writer should be either an engineering major with writing skill or an English major with an engineering background has not yet been clearly established, although it appears that the former has the wider support in industry.

The potential writer is faced with the dilemma of deciding what branch of engineering to undertake. Majoring in a mechanical engineering discipline would likely eliminate any chance of employment in the electronics or medical field and viceversa.

Although there are only a dozen or so educational institutions in North America and Canada offering either a four-year or undergraduate degree program in technical writing, none of these are able to augment the writing portion of the course with the general engineering degree program that is so necessary for the technical writer to have; such a program would enable the writer to make an early assessment of the particular field of endeavor suitable for his or her talents.

At this time, especially in areas where no formal training programs exist *per se*, the technical writer appears to have essentially materialized into the profession by two possible methods:

a. Experienced engineers and technicians who, with some high level of writing talent, have taken up the pen with the intention of continuing in this field. Their writing ability has been upgraded with extension courses or part-time study at a university toward an undergraduate degree in English.

b. The English undergraduate, or even high school graduate and who has, by means of in-plant training, seminars, extension courses and so on, been able to gain sufficient technical familiarity to function as a competent writer in a field that may be somewhat limited, for example, software (computer programs) or legal writing to name just a couple.

EMPLOYMENT OPPORTUNITIES

There has been a marked upsurge in the demand for technical writers in the 1990s and the opportunities for those wishing to enter the field are limited only by their willingness to take advantage of whatever training courses and programs are available, and by industry's willingness to take a chance and engage writers that have little or no experience in their specific field of operation.

CHAPTER 3

TECHNICAL MANUALS AND HANDBOOKS

Comprehensive instruction manuals are needed for special-to-type commercial equipment and products that are sold directly by manufacturers, using either their own sales organization or exclusive sales-representative organizations.

Most commercial products, particularly those that require detailed technical instruction manuals, are usually sold on special order to government agencies or large organizations who undertake the maintenance of the equipment themselves. Such equipment is not manufactured or assembled until the customer specifies all the options and establishes the performance requirements of the specific model required. Manuals for such equipment may be organized and written for various basic models, options, and combinations of such options.

These instruction manual sets can usually be prepared in large quantities, especially since it is likely that a sufficiently large number of units will be sold to justify the printing of a fairly large number of copies. Because the basic design of the equipment itself allows little or no variation in the product itself, other than that permitted by the manufacturer, the manual can be written to cover all variations, major as well as minor. Manual sets can then be assembled from bulk stock to meet any specific order.

The manual is usually packed or shipped with the equipment. In some cases, advance copies of manuals may be required for in-house training of those personnel who will ultimately be responsible for the day-to-day maintenance of the equipment. Where this training requirement is a regular occurrence, manuals of lesser complexity may be prepared.

These would generally be devoid of such information as parts listings or major overhaul details and would essentially be technical information and operating procedures.

TYPES OF MANUALS

General Manuals

The one-of-a-kind equipment that requires technical manuals will be found almost exclusively in the commercial marketplace. This equipment may be of two types: those that are specially designed for a specific application and require original engineering not likely to be duplicated in another order; and those that have so many options that it would be unlikely that two machines or systems would be put together in the same way, even though they may bear the same model designation.

Such specially manufactured equipment could be a police vehicle communications set, a sophisticated alarm system for a business or home, an industrial power turbine generator for backup power, a small airport weather radar system, or a mainframe computer. These have such a high initial cost (per unit or as a large order) that the cost of a specially written manual is but a very small fraction of the total contract price. It is for these applications that custom manuals are usually produced.

Manuals for the second type of equipment can often be handled in the same manner as those for off-the-shelf items. The equipment may be "individualized" and distinctive such that some components or even full operating sections are different from other components or sections in related and similar equipment. A prime example is the IBM-clone personal computers on the market today. A specific model may vary in its makeup by having different hard drive sizes and types, various quality interface boards, plus any of the multitude of optional add-ons available.

Instruction manuals for such equipment are prepared in the same manner and appear similar to manuals prepared for large stock orders of standard and identical equipment. A simple statement or specification sheet can be inserted in the text where information is to be given about each distinctive feature. This statement will tell readers to ignore the instructions for a component or section unless their equipment contains the particular component being referred to.

On the other hand, it is possible that the operating procedures will differ, depending on the combinations of units purchased. In this case, one discussion would be included to describe one mode of operation, with an appropriate introductory statement, and other discussions would describe the other modes of operation. A statement describing each mode and appropriate instructions to be followed would precede each discussion.

Since the manufacturer has to budget labor and money to pay for manuals, he does not wish to purchase more writing labor in place of more printing costs. Printing additional quantities is cheaper, by far, than writing many versions of essentially the same material. Also, the problem of storage and selection of the correct manual for shipment to the customer add to the warehouse and labor costs. Therefore, whenever possible, all information concerning a class of equipment should be included within a single manual.

Custom-Made Manuals

Occasionally it is necessary to provide a special manual for a single item of equipment. Writing procedures to be followed are similar to, but differ in the degree of effort from, those utilized in writing manuals for equipment with a large number of individual subunits. For one thing, purchasers of the equipment are already aware that they are procuring a special product.

Considerable negotiations concerning the technical aspects and the overall cost have preceded his purchase. He should also have been informed by the sales department that the standard manual will not fully cover his equipment and that he must either procure a custom-made manual or accept this standard manual and make his own revisions to it.

Assuming that the buyer does not wish to be involved in the amendment of a standard manual, and is aware of the special nature of his purchase, what steps are available to him to ensure that the technical information he received is adequate to cover all aspects of the equipment? The customer must still have a comprehensive manual but may not need one of the quality or content of the standard manual. He can receive a manual that has fewer photographs—especially since the most applicable photographs would be of his own equipment.

He may get fewer specific operating instructions, especially since he has modified the controls to meet his own needs. He may get less maintenance information and possibly less information pertaining to lubrication and other periodic servicing.

However, he would still get all information that pertains to the type of operation, rather than the specific methods of operation and control, used by the operators. More than likely the operating instructions will have to be abbreviated. It is also possible that there may be a change in the maintenance instructions.

There are some types of equipment that cannot do a single useful operation except when a combination of subunits are assembled within it in a specific and prescribed manner, such as a system. The only similarities between different equipment of this type are in the individual components. This type of equipment is found in ships, aircraft, airport air-traffic-control centers and so on.

This approach works successfully when the equipment operation is virtually automatic and the manual covers maintenance instructions almost to the exclusion of operating instructions. Since there are no modes of operation to be concerned with, it is necessary only to ensure that the equipment functions in accordance with its performance criteria.

Either it functions correctly or it does not! If it does not, it requires servicing, and the manual should adequately cover the requirement.

Custom manuals may be produced by combining special instructions for each of the subunits or the manual may be written from scratch to cover a specific assembly of subunits. This latter approach is expensive and time consuming and is usually applied only in large quantity purchases by governments and the like.

ARRANGEMENT OF TECHNICAL MANUALS

The more common arrangement of a technical manual can be broken down into six major parts:

1. General Information

2. Operating Instructions

3. Theory of Operation

4. Maintenance Instructions

5. Parts List

6. Installation Instructions

The actual titles used for each section may differ slightly in accordance with the manufacturer's preferences, but the content should essentially be the same.

General Information

This section can contain quite a lot of technical information under various headings. The information may be only one or up to a dozen paragraphs. These subheadings could be:

Specifications—a page or two describing the equipment in a brief form similar to that of a data sheet. It would list, for example, its reliability figures (MTTR), some characteristics of its operation, controls provided, weights and sizes, and power requirements. The following sections, however, may provide the same information in more detail. The use of this information sheet would be determined during the manual planning stage.

Introduction—may introduce the manual and its purpose.

System Instruction Manuals—describing the arrangement of the manual set if it is relatively complex. It may refer to other manuals that also relate to the equipment but are not issued with this set.

Purpose of the Equipment—simply states the function of the equipment and perhaps its state-of-the-art technology.

Description of the Equipment—briefly describes its physical characteristics: how big it is, what it needs to operate (power), how many major parts go to make it up, where and how it would be installed.

Operating Principles—a basic description of the way in which it functions. Very little detail, but enough to give a layperson an idea of its operation.

Equipment Configurations—if the equipment is one in which various configurations are possible, these would be listed with a brief comment about each (e.g., a high- or a low-power system, different testing/monitoring systems, and options available). This subsection may be followed by a description of all the subassemblies comprising the system. The decision as to the depth of technical information within this section is one that would be made during the manual planning stage.

Operating Instructions

This is simply what the title suggests and would include all modes and methods of operation, plus the results expected from each type of operation or control. It would include a summary of controls and indicators, augmented by either line drawings or photographs of the position of each. If these controls are numerous, tables could be used to list each subassembly with the control reference number, its name or marking, and its function.

If the equipment is sufficiently complex, there may need to be a detailed operating technique available for the operators. Generally, short training courses are provided for operating personnel, by the manufacturer, if the equipment is user-intensive. The inclusion of this operator training material in the manual may be unnecessary, but if provided as a separate manual, it would need to be referred to in the System Instruction Manuals section of the General Information.

Theory of Operation

Once equipment goes into the hands of the user, it serves its function only as long as it is performing correctly. For most commercial purchaserbs, the operation of the equipment is critical to their profit picture, as good operation means a better profit-and-loss statement.

There are commercial operators who operate equipment until it requires extensive maintenance, even repair and overhaul. This practice is justified on the grounds that they buy the best equipment available, usually equipment that is extensively overdesigned. They claim that the cost of lost production as a result of periodic preventative maintenance plus the maintenance cost itself is greater than the cost of lost production plus the time and cost of emergency repairs. To meet the needs for such operating practices, only good operating instruction manuals are needed, at least at the onset.

Maintenance Instructions

Maintenance must be considered a separate function from operation. It is performed by personnel whose skills differ from those operating the equipment. Maintenance may be performed on a test bench, in a fixed location, in a specialized service department, or partly at the operating position and partly in the maintenance shop.

The location at which maintenance is performed is seldom of concern to the manufacturer of the product or the writer of the manual (except where the manufacturer has to provide specialized tools or test equipment or if special conditions such as "air-conditioned" or "clean" rooms are required). The manufacturer need only specify the special tools and equipment that are needed to perform the maintenance function and how to verify the quality of the work performed.

In some instances, the specialized test equipment is sold as part of the contract or is merely specified as necessary for the maintenance program.

For practical purposes, the terms *service*, *repair*, and *overhaul* describe different phases of basically similar functions, varying only in the degree of effort and the expertise of the personnel who perform the maintenance operations.

Here again the particular industry purchasing the equipment provides guidance as to the methods of organizing the instructional manual. While military customers have rigid rules and specifications for manuals prepared for their equipment, most nongovernmental industries prefer to have all instructions pertaining the nonoperating aspects of the equipment in a single, easily referred to volume.

This single volume would cover all periodic service and inspection, lubrication, replacement procedures of a special nature, complete listing of parts, disassembly procedures, drawings showing the components of each major - assembly, and the special maintenance conditions that may prevail.

In order to verify the quality of maintenance, most maintenance manuals include some procedures, within either the maintenance manual or the operating manual, which can be used to test the performance of the equipment.

The sublevels of maintenance instructions can be explained as follows:

a. *Field Service:* These are instructions for making emergency repairs of a minor nature at the operating site. Such instructions call for few and simple tools. They also provide procedures to be followed to ensure that no further breakdowns occur or to prevent equipment from being used if a breakdown is imminent. This type of minor servicing may be performed by operators under certain conditions.

Most equipment being designed today is fitted with built-in test equipment (referred to as BITE). These BITE units can range from simple switchable test functions that locate already-existing problems by search-and-locate operations to more sophisticated microprocessor-based systems that can constantly monitor and test the equipment's operation and, if a potential problem is found, act accordingly. These BITE units would, after raising the appropriate alarms, either shut the system down, indicate the problem and wait, or transfer operation to a serviceable standby.

Irrespective of how the equipment problems are diagnosed, the servicing procedures for the system at the field repair level would not be unduly complex. However, the advent of the BITE requires a new manual section covering the operation and maintenance of this specialized test equipment.

b. *Field-Shop Service:* These instructions would apply to a fully equipped service center. In a factory with fixed equipment, it would involve service on a machine that is fixed in place and would cause the rescheduling of work that is normally performed by this machine.

c. *Depot Repair:* If the nature of the maintenance requires yet more specialized and complex tools, as well as calling for more specialized knowledge and maintenance skills than is available at the field-shop level, the manual for this level of service must be yet more complete. This level is comparable to the specialized services provided at automobile-body shops, electrical shops, electronic service centers and so on. In the industrial plant it would involve the separate specialities of the electrical shop, tool shop, hydraulic shop, and so on.

d. *Overhaul:* This type of work is comparable to the complete rebuilding of a machine. In practice it is performed by skilled specialists who seldom need the manual except for reference. Such personnel possess skills that frequently match or exceed the skills of the original equipment manufacturer. At this level, major parts may be replaced or rebuilt.

Unless all material is contained in a single manual, no manual of instruction will completely cover all levels of maintenance as each company may define them. Even the most complete manual cannot cover every possible arrangement and maintenance condition that may occur.

The skills of the operators and the servicing personnel must be taken into account. Obviously an equipment is designed to function as flawlessly as possible; however, this rarely happens. To provide any more than basic instructions for normal operation and maintenance procedures is not feasible and is unnecessary under such circumstances. Field personnel will find that they will encounter certain problems with a fair degree of consistency, and they will eventually learn to cope with these problems with a minimum of disruption to the operation of the equipment.

Occasionally, equipment, particularly a production machine, will be transhipped from one part of an organization to another or even to a different company. This is common practice in multiplant companies where less sophisticated machinery will still function adequately to meet the needs of the local market. The newer machinery will be used to replace the older machines in more competitive market areas.

When such equipment is relocated, it is first dismantled, marked, shipped, and then reassembled at its new location. However, since it is already an old machine, it will likely have suffered from considerable use and before being reassembled at the new location it may be subjected to a complete overhaul. This means the replacement of many parts that are still functional but may prove unreliable after further extended use.

In other industries, equipment is used so extensively that it requires an overhaul at relatively frequent and specific time intervals. This is true of construction, automotive, and aircraft equipment, railroad stock, and other machinery that is operated for long periods of time at locations away from a maintenance depot. These major equipment items are generally rotated in the field to enable maintenance and overhaul to be performed at some predetermined interval.

Reliability of operation is essential to profit-making operations. Most commercial equipment is given a Mean Time Between Failure (MTBF) calculation, which signifies the time before an expected failure occurs. This figure is the result of a complex calculation, based on the expected life of each component in the equipment. Periodic overhaul is one way of ensuring that this figure is either maintained or improved. A simple example is a pulley belt on an electric motor-driven machine. If it is rubber or some similar material that provides flexibility, it has a definite life, that is, a period during which it will maintain its flexibility and strength. After that time it should be replaced to ensure continued operation of the motor. This replacement before breakage is termed *preventative maintenance*.

When industrial or commercial practice involves established overhaul procedures, it is feasible and often desirable to write separate overhaul manuals, describing all procedures specifically appropriate to the overhaul function.

Spare Parts

Some equipment is used within industries that have a fairly complete maintenance organization, with quality tools and many types of commercial hardware and other parts on hand or easily obtainable. In some instances, spares have been obtained from the original equipment manufacturer as spares packages. These customers therefore want to have a spare parts breakdown that not only describes, in detail, the special parts but also standard commercial hardware that is approved for use within the equipment. Figure 14-1 shows an example of such a parts list, together with its associated exploded-view illustration in Figure 14-2. (Refer to Chapter 14, *Illustrated Parts Breakdown* for a more complete discussion on this subject).

Installation

This chapter or section covers the setting up or installation of the equipment to the point where it is ready for use. It may involve reference to the floor plan of the machine, unwrapping and special wiring for an electronic device, or the installation of a component. The operator may be able to perform the installation alone or may have to be assisted by other persons. For instance, the operator may be able to unpack the electronic equipment but may not be able to perform the initial performance and calibration checks.

If electrical hookups have to be made, adequate wiring diagrams must be included. It must be complete enough for the purpose but should not include internal component wiring such as would be needed for trouble-shooting; that is included elsewhere in the manual.

The installation section may include illustrations of the equipment in various stages of assembly and pictures of each individual component needed to assemble the equipment. There would most likely be an itemized checklist of parts and functions. There may be instructions for unwrapping if specific routines have to be followed. In some instances there will be instructions for the disposal or retention of wrapping, crates, and so on, if reshipment is a possibility.

Because an installation section is used, at the most, perhaps twice during the lifetime of the equipment, it is worthwhile to consider locating it at the rear of the manual in an appendix. This way, the front of the manual is not cluttered with many pages of information that is irrelevant to the day-to-day use of the manual.

Also, depending on the binder type, the whole installation section can be removed and put into a smaller, more convenient binder for use by the installation group. This protects the master manual from being damaged if accidentally left on floors or in open areas during construction and installation work.

In some instances the installation instructions, because of their complexity and magnitude, have had to be prepared in a separate manual (e.g., the installation of a major aircraft navigation or radar system at an airport, where the work involves the erection of large towers, antenna arrays, and a complex calibration stage with an orbiting aircraft).

CHAPTER 4

PLANNING A TECHNICAL MANUAL

INITIAL DECISIONS

When deciding what form a manual will take and how to tackle the task of preparing it, certain decisions must be made. The first decision is to determine audience or end use: Is it to be a maintenance manual, an operators manual, or for use as both? Once that decision has been firmly established, the structure of the manual can be decided. This is followed by the preparation of a plan that will specify the steps and procedures necessary to combine all the technical text and illustrations of the manual into a finished product.

DEFINITIONS

The following are manual terms used within this book.

System: Generally used in conjunction with a major item of equipment: for example, a data acquisition system (computers), an instrument landing system (navigation aid), a fuel-oil handling system (thermal power plants), or a desktop publishing system.

Subsystem: Essentially a major element of a system. The printer unit of the desktop publishing system be considered as a sub-system.

Part: A main division within the system technical manual; this may cover a subsystem, such as the printer unit, or a specific subject, such as a repair manual or an operating manual.

Volume: A subdivision of a manual or a part thereof providing handling convenience only.

Section: Contains material relating exclusively to a subsystem. Essentially a subdivision of a part.

Chapter: A collection of paragraphs and other text covering a main subject within each section.

Paragraph: Self-explanatory. The paragraph should not be further subdivided beyond sub-subparagraph level.

Appendix: A document containing material that is relevant to the material within the manual but cannot or should not be included within the body of the text. Appendixes are always located at the rear of the manual (this subject is covered further in Chapter 15, *Appendixes and Addenda*).

A block diagram of a basic manual structure is shown in Figure 4-1. Note that the manual breaks down into a number of subsystems. Each subsystem then breaks down into sections, which in turn are broken down into chapters. Depending on the complexity of the equipment, the manual may be structured from either the system, subsystem, or section level down.

MANUAL STRUCTURE

In today's world of high technology, technical manuals for electronic/ electrical equipment, for instance, generally require a greater level of documentation and detail than mechanical or electrical equipment and also require a greater level of user expertise. The percentage of electronic computers and controllers installed in the modern motor vehicle is to the point where the 1960s style motor mechanic has been replaced by a mechanically orientated electronics technician.

To illustrate the variety of manual forms we can consider four examples of electronic equipment that use different forms of documentation.

Example 1: Consider a piece of equipment such as a personal computer that uses many subassemblies from OEM (Original Equipment Manufacturers) sources, that is, modules built by other manufacturers and sold in bulk to be included in other manufactured items. A computer manufacturer may use 10 or 12 major items and subassemblies.

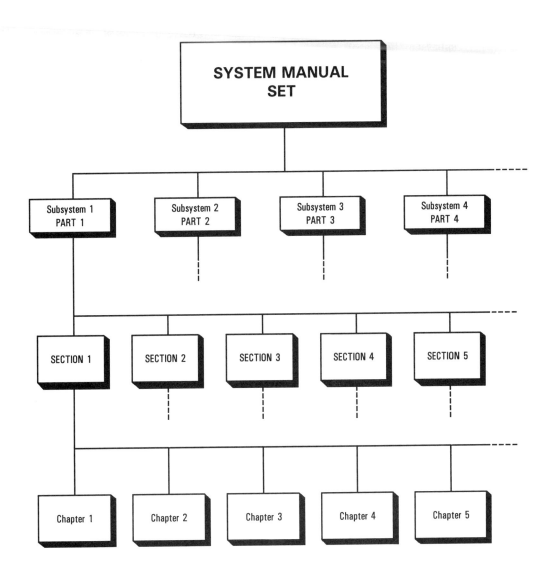

FIGURE 4-1
Basic Manual Structure

Because of the nature of computer manufacture and marketing today, little or no technical information will normally be provided for most of the subassemblies, apart from a brief summary of specifications. As each subassembly is small and often purchased from different OEM sources during the course of the computer's manufacture, it is neither possible nor feasible to provide technical information to the customer. The owner's manual may contain only relevant setting up instructions, operating instructions, a technical summary or specification page and some type of very basic troubleshooting chart.

Generally, the software programs supplied with the equipment will be adequately covered with the applicable manufacturer's handbooks.

This style of manual is prevalent with equipment provided to the "domestic" market. The user is invariably a nontechnical person and any form of service is usually undertaken by the dealer or a third party such as a general computer repair facility. Even in such a facility, the level of repair for most smaller subassemblies is generally on the basis of a line replaceable unit (LRU), which may be a printed circuit board (PCB) or module. The unserviceable item is simply thrown away and a new or overhauled one installed. Some high-value or major items may be considered economically repairable and would be returned to the manufacturer for repair.

Example 2: This example of manual preparation concerns an OEM supplied subassembly such as a power supply unit, which forms part of a large, very expensive mainframe computer or similar equipment. As maintenance on a regular basis and the ability for localized repair are essential, the documentation must be extensive. If not, some arrangement must be made to have spare units on hand and the provision for return to the manufacturer or other facility for repair.

In the former case, the manual set for the main equipment should include all the necessary documentation for servicing this subassembly. This information would be integrated into the manual and should comprise the theory of operation, repair/overhaul instruction, and a parts list.

Example 3: This concerns the manual type for radio-telephone equipment (i.e., a two-way radio set) in a transit vehicle. In this instance, the manual would likely be structured from the subsystem level down. Hence the sections would comprise the following: transmitter, receiver, power supply; control unit, and antenna system. The chapters would then cover technical description, repair, operation, parts, and so on. There would be an introductory chapter covering the unit as a whole and some general operational procedures, installation, and recommendations.

It is relatively easy to see that these three manuals would require arrangement of nothing more than a single volume divided into sections and chapters.

Example 4: We now look at the preparation of manuals for large, complex types of equipment. The example to be used is an item of electronic equipment, namely, a ground-based radio navigation aid used for air traffic control at commercial and military airports throughout the world: an Instrument Landing System (ILS), which, because of its complexity, requires a very comprehensive set of manuals. Using this as our example enables us to cover all possible aspects of manual preparation.

Essentially, an ILS installation comprises two different electronic systems (one *Localizer* and one *Glidepath*), two different antenna arrays, and one common OEM-supplied emergency power supply equipment. As the two systems are located at different parts of an airfield, there is a need for two separate sets of manuals. In some cases, in systems of this nature, a third manual covering the whole integrated system may be necessary.

The manual for each system will be divided into a series of parts, each being a self-contained subsystem manual (a similar approach is taken in vehicle service manuals—a chapter each for the engine, emission controls, fuel system, brakes, drive train, body, electrics/instruments, etc.).

Before going any further, it must be reiterated that the basic and fundamental use of a technical manual is to provide every possible item of information that will aid in understanding the operation of the equipment. Adequate technical information must be provided so that the operator can use it competently and so that efficient and thorough repair and/or maintenance can be carried out. However, providing this information in a haphazard and ill-conceived form negates any value the technical material has.

Another very important aspect to consider with manuals of this type is their use as user-training material. With equipment as complex as an ILS system, technician training is almost as important as the actual provision of the technical information. Invariably the manual serves as the main textbook for the training of the technical staff, who will eventually use the manuals in the everyday operation and maintenance of the equipment.

With these factors in mind, it is possible to prepare an outline of a typical technical manual:

Front Matter:

> Warranty Information
> Table of Contents
> First Aid/Rescue
> Hazardous Material Handling (if any)
> Illustrations of the System

Part 1—System Information

Specification Sheet: A brief summary of the equipment, including physical dimensions and weights, electrical requirements, and operating and performance specifications (limited to about two pages).

General Information: The purpose of the system, detailed technical specifications, operating characteristics, equipment supplied, additional equipment required, plus a list of all possible options.

Part 2—Operating Instructions

> The procedures for operating the equipment should be detailed in this part. It must contain all information necessary for nontechnical staff to understand the operation of the equipment if it is operator-heavy rather than an item of equipment used solely by technical personnel. When the operating procedures or the equipment is complex, it should be arranged as an on-the-job-training procedure, complete with illustrations.

Part 3—Theory of Operation

> This part of the manual is where the user is given information about, and hopefully understands, the technical operation of the equipment. An explanation of the operation of each major subassembly should come first and utilize functional block diagrams only.

Part 4—Test and Alignment

This part can also be called "Maintenance" and would contain test equipment data, preventative maintenance, removal and replacement procedures, trouble analysis information, and adjustment procedures. It would also include appropriate illustrations of test setups and adjustment procedures.

Part 5—Parts List

This part of the manual would contain an illustrated parts breakdown (IPB) of the main and subassemblies with a list of referenced parts, part numbers in numerical order, and a reference designation index. It could also contain a separate list of items considered to be "consumable" if applicable.

Part 6—Drawings

A separate area for all detailed schematic, wiring, and mechanical drawings. The block diagrams or the illustrated parts breakdown diagrams would not be repeated here. Figure numbers would be referenced throughout the text and the parts listings.

Part 7—Installation

If the equipment is, for example, a relatively simple item (e.g., a depth finder for a boat), the installation could be located in Part 1. If the installation is very complex, sometimes needing concrete foundations or special erection equipment and procedures, the installation instructions should have their own section at the rear or be set out in an appendix.

Figure 4-2 is a typical manual structure chart for a large and complex item of equipment. By reference to this structure chart, it will be possible to divide up the manual material into blocks and then determine the most suitable layout.

MANUAL PRODUCTION PLAN

Once it has been determined how the manual is to be made up, that is, how many parts, sections, chapters, and so on, it will consist of, it will be necessary to formulate a production plan. This is, in effect, a form of diagram showing the flow of material, information, and activities and includes critical dates if the manual production is tied to a schedule. We can utilize either a simplified block diagram style or a more sophisticated project management system such as *Project Evaluation and Review Technique* (PERT) or *Critical Path Monitoring* (CPM).

Using either of these systems will provide an immediate view of the needs of the publication program and should ensure that no important factor is overlooked, which could cause serious delays to the production of the manual.

Figure 4-3 shows the block diagram form of production plan, which is less concerned with dates than with what activities have to be performed. This is suitable for jobs that do not have a strict deadline for completion.

The CPM flowchart or network diagram shown in Figure 4-4 is an activity orientated plan, which, when used properly, can keep strict control of the manual preparation activities and ensure that specified completion dates will be reached.

While it is not the intent of this book to provide training in PERT or CPM, the following brief explanation of the activities in a simple network diagram may indicate the value of this program to a company and initiate some further reading on the subject.

The main or *critical* path is shown with double arrows, $\Rightarrow$, with the secondary paths shown as $\rightarrow$. The circles or *nodes* are divided so that the activity number is in the bottom half, the *earliest* date the activity can start is in the upper left portion and the upper right portion, has the *latest* start date.

Node numbering can start at 1 and use single digits, but by using 10's it is possible to back up the program or insert several supplementary networks at any time to take care of any unforeseen activities.

In Figure 4-4, the following activities can clearly be determined, beginning at node 10:

a.   The text writing and/or assembly of material must begin (path 10–20),

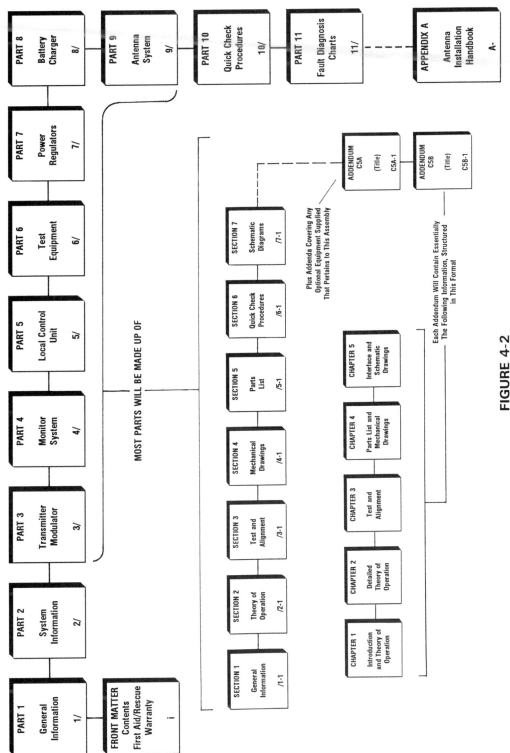

FIGURE 4-2
Manual Structure Chart

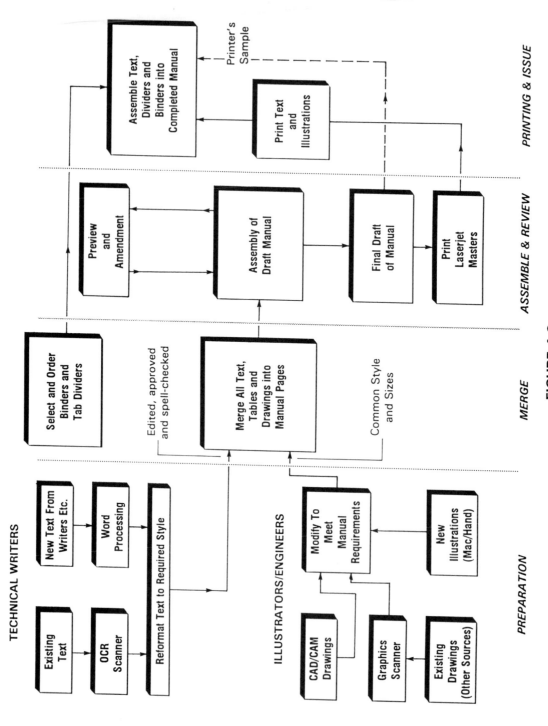

FIGURE 4-3
Production Plan – Block Diagram Style

b. Illustrations must be started and be ready in draft form by node 40.

c. The writing or collation of appendix material must be started (path 10–12).

d. Both binders and tab dividers should be ordered (paths 10–90 and 10–80).

e. The design of the slip-in manual front cover sheets and spine strip should be started (path 10–11).

f. All front matter material should be started (path 10–50).

Note that the paths of the illustrations and the text, between 10 and 40, could be interchanged if the illustrations were of major significance. This would be the case where the illustrations were complex and expected to take longer to prepare than the text.

The next significant event is when the edited and formatted text plus the illustrations are merged together (node 40) to produce the first draft of the manual. From this point, the table of contents can be prepared as page numbers should be available.

The first draft is prepared between activities 40 and 50, then merged with the already finished front matter. Activity 50–60 will be the review of the draft manual and will include any revisions needed. There may be several cycles of the review and revision process here. At the completion of this activity, the appendixes, if any, will have been independently prepared, reviewed, and amended if necessary and will be ready for the next activity (60–70), which will be the preparation of the manual final draft as a camera-ready document. Note that at this time, the table of contents (40–60) should have been prepared and should be ready for insertion into the final draft.

After the printing process (70–80), the tabbed dividers must be added to the material so that the manual can be collated ready for insertion within the binders. At activity 90, the bundles of manual pages, binders, printed cover sheets, and spine strips are all brought together for the assembly process (90–100).

Cumulative timing (hours, days, or weeks as appropriate) is normally inserted in each pathway between the nodes with the critical dates within the node. As the job progresses, the flowchart can be updated on a regular basis to ensure that the program is being maintained on schedule.

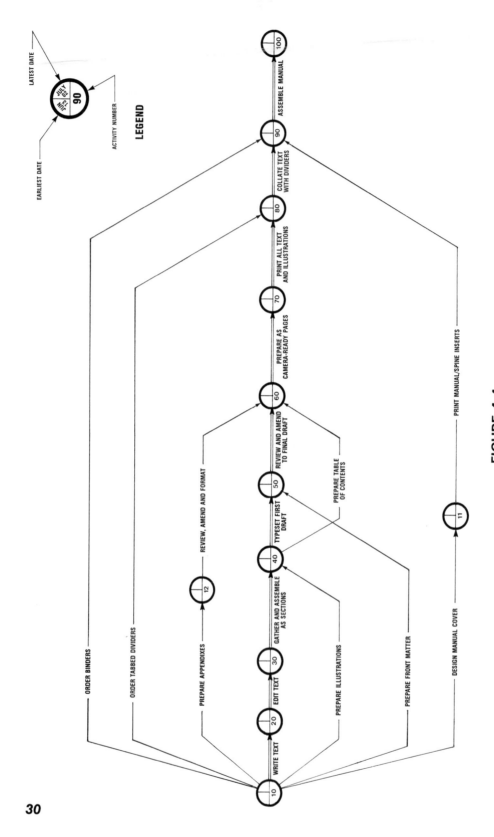

FIGURE 4-4
Critical Path Monitoring

LEGEND

LATEST DATE

EARLIEST DATE

ACTIVITY NUMBER

ORDER BINDERS

ORDER TABBED DIVIDERS

PREPARE APPENDIXES

REVIEW, AMEND AND FORMAT

WRITE TEXT

EDIT TEXT

GATHER AND ASSEMBLE AS SECTIONS

TYPESET FIRST DRAFT

REVIEW AND AMEND TO FINAL DRAFT

PREPARE AS CAMERA-READY PAGES

PRINT ALL TEXT AND ILLUSTRATIONS

COLLATE TEXT WITH DIVIDERS

ASSEMBLE MANUAL

PREPARE TABLE OF CONTENTS

PREPARE ILLUSTRATIONS

PREPARE FRONT MATTER

DESIGN MANUAL COVER

PRINT MANUAL/SPINE INSERTS

CHAPTER 5

PUBLISHING SYSTEMS

Deciding on a method of publishing an in-house manual for the first time is often difficult in view of the variety of modern word-processing programs, desktop publishing systems, and software packages available. By examining several different manual preparation systems, it should be possible to decide on which route is to be taken.

The parameters on which the choice is made are basically those relating to the company's long-term investment in manual production and/or upkeep, and the way in which the manual is to be prepared. Essentially, there are about three basic methods generally available for the in-house preparation of a complete manual or the preparation of camera-ready copy for an external printer:

a. Typesetting with a sophisticated desktop publishing system that uses special workstation hardware and software. Typical examples are **Apollo** or **SUN** workstations fitted with either the *Interleaf 5* or *Zerox DocuBuild* software. Programs such as these easily handle manuals containing up to 1000 pages, constructing and displaying every page in WYSIWYG* form, complete with line art, photographs, and tables. Virtually every typesetting and layout feature used in modern publishing is included plus some significant features such as the facility to provide immediate rearrangement of the document following a revision (table of contents, page numbers, page readjustments, and even marking of affected text/paragraphs) so that the document-in-progress is always at the latest revision level.

---

* A common acronym in electronic publishing meaning "what you see is what you get." The screen will display all text, typefaces, margins, columns, and so on exactly as they will appear when printed.

These publishing systems have high initial equipment and software costs as well as high maintenance costs for both and require specialized training for the operators. They are generally found in professional publications organizations and companies who are continually preparing and updating large volumes of technical documentation.

b. Typesetting on a smaller scale with specialized computer systems such as an **Apple Macintosh** fitted with *Quark XPress* (text handling) and *Adobe Illustrator* (illustrations). The Macintosh system lacks some of the sophisticated features of the Interleaf and Zerox systems, inasmuch as two separate software packages have to be used to combine text and illustrations, but of most importance, initial and upkeep costs are minimal.

As with any of these systems, well-trained illustrators and writers are needed to complement the operators and ensure that the material being inserted into the manual is of the necessary standard. For a medium-sized manual of about 100 pages, the Macintosh is an excellent choice.

c. The final system to be considered is probably the most economical and most widely used. This is the common office-type personal computer fitted with a low-priced desktop publishing program such as *Ventura Publisher* or *Aldus Pagemaker*, which are relatively sophisticated writing, design, and production programs. These programs do need a certain amount of specialized training to take full advantage of their features, but the overall cost is still very low.

Alternatively, the personal computer can be fitted with a word-processing package such as *Microsoft Word* or *WordPerfect*. These programs, used in conjunction with add-on typeface packages, clip-art, and other graphics packages can be used to produce excellent manuals.

The biggest drawback with these packages is their limitation in processing more than several pages at a time. While they have the facilities to be able to produce very creative layouts, their inability to create a lot of page layout functions automatically is a disadvantage; however, for small manuals, being prepared to a minimum budget, this method is unsurpassed. Also, short familiarization courses and on-the-job training are usually sufficient operator training for these programs.

CHAPTER 6 ──────────────

LAYOUT AND FORMAT

INTRODUCTION

Layout is a term used to express the general appearance of a manual with respect to page form or structure, that is, the positioning of the text, tables, and illustrations, the type and size of text used, the style of the headings, page numbering, and so on.

By adhering to some simple rules, it should be possible to develop a suitable design for the manual that provides the best possible presentation of the technical material within it, and one that reflects a high degree of professionalism within the company.

Some sections of the manual, namely, the parts lists and the maintenance section, will undoubtedly require a different approach because the material is generally sequential rather than narrative.

TYPEFACES AND TYPESTYLES

The art of selecting and arranging the various forms of typefaces to form a harmonious and effective display in a manual layout is known as *typography*. Typography for technical manuals is relatively simple compared with the requirements of more sophisticated "display" types of documents, such as brochures, and annual reports.

Legibility

Legibility is the quality of writing that makes it possible to read. *Readability* is the characteristic of a body of type that makes it comfortable to read.

For most technical manuals, the selection of a typeface is relatively simple—it is usually either basic Courier, the more professional Times Roman, or Helvetica. The last two generally differ only in point sizes, styles, and weights and provide enough variation to fulfill most requirements of this type of publication.

Readability

Readability is relative to the reading ability of the manual's user. Within the limits of normal manual use, that is, with adult readers, minor differences in type size are not significant.

The readability of a page is affected by no less than eight factors:

1. Typeface (character of the typeface).

2. Size of the typeface.

3. Leading (line-to-line spacing).

4. Length of lines of text.

5. Page layout (column widths and numbers, margins, figure and table locations, etc.).

6. Contrast of type and paper (which includes color).

7. Texture of paper.

8. Typographic relationships (headers, page numbers, etc.).

Some factors are more significant than others, but it is their combined effect that gives the page its character, and it is only when all are in perfect balance that a truly readable page results.

Typographic Terminology

To further understand the basics of typography, certain terms used within publishing and typesetting should be explained:

Font: a complete assortment of any one size and style of type containing all the characters for setting ordinary composition.

Typeface: The character set in one size and style. It may be either of the following:

>   **Serif**: Generally found in lettering of the Roman typeface. A classic style with embellishments at the edges. Common types are Times Roman, Dutch, Schoolbook, and Baskerville.
>
>   This typeface can be found in medium, medium-italic, bold, and bold-italic styles.
>
>   ABCDEFGHIJKLMNOPQRSTUVWXYZ
>   abcdefghijklmnopqrstuvwxyz
>   1234567890 &?!$(.,:; " " ' ")
>
>   **Sans Serif**: Lettering based on a mechanical form, no serifs, and generally in the form of straight consistent strokes. The most common types are Helvetica, Univers, and Swiss.
>
>   This typeface can be found in a variety of styles; light, regular, medium, bold, and condensed. Each of these can also be found in an italic version.
>
>   ABCDEFGHIJKLMNOPQRSTUVWXYZ
>   abcdefghijklmnopqrstuvwxyz
>   1234567890 &?!$(.,:; " ')

Pica: A pica is a *linear* unit of measurement used principally in type-setting and publishing. There are six picas to the inch, each one being divided up into 12 points. Pica rules are readily available at stationers.

Point: Another unit of measurement, used principally for designating type sizes. There are 12 points in a pica and 72 points to an inch.

Point Size: The size of a typeface is referred to in points, that is, its height from the upper to the lower limit (top of a d to the bottom of a y) is measured in points, for example, 30-point Times Roman.

Leading: Pronounced *ledding*, a term derived from the word lead, because in the old hot slug-type composition days, thin metal spacers were placed between lines of type to allow for extra white space. This term now refers universally to the distance between successive rows of text. If no extra white space or leading is inserted, the type is said to be *set solid.*

The following is an example of 7-point type, set solid:

> The amount of space between lines is known as leading. There is no set rule to follow. Too much leading can often be as bad as not enough. A point or two of leading will vastly improve the readability of the text.

However, extra white space gives better appearance and increases readability, so leading is usually added. The measurements are made in points from baseline to baseline. Hence 7-point type set with 2 points additional space between lines would be described as either 7-point on 9-point (7 on 9, 7/9) or 7-point, 2-point leaded.

This is the same sample block of text set with 2 points of leading:

> The amount of space between lines is known as leading. There is no set rule to follow. Too much leading can often be as bad as not enough. A point or two of leading will vastly improve the readability of the text.

The optimum leading for this example is 7/8 or 7-point on 8-point (1 point of leading) as shown below.

> The amount of space between lines is known as leading. There is no set rule to follow. Too much leading can often be as bad as not enough. A point or two of leading will vastly improve the readability of the text.

Kerning: Refers to the adjustment of space between selected pairs or groups of letters. Although most computerized typesetting and printing equipment is able to perform an automatic proportional spacing function to proportion the space allotted to letters of individual words, it is a limited function.

It is, however, the kerning function that *finishes* an otherwise loose line of text by the fine adjustment of each individual word and its surrounding space to correctly proportion the line of text.

The following samples of kerned and unkerned text show this variance:

A block of 10-point Times Roman, leaded 2 points and set without any kerning.

Times Roman is one of the most versatile and
popular faces ever designed. It improves so much
on the imperfections of the romans that preceded it
that it is almost impersonal in appearance, although
it does seem to acquire the personality of

The same block of text set with one unit of kerning. The setting is notice-
ably tighter, but still retains good readability.

Times Roman is one of the most versatile and popular
faces ever designed. It improves so much on the
imperfections of the romans that preceded it that it is
almost impersonal in appearance, although it does
seem to acquire the personality of whatever subject is

Same block again but set with two units of kerning. Setting is much tighter
than normal and this is about the limit that would be used before characters
would overlap and severely impair readability.

Times Roman is one of the most versatile and popular
faces ever designed. It improves so much on the imperfections of the romans that preceded it that it is almost
impersonal in appearance, although it does seem to
acquire the personality of whatever subject is

Readability: There are few aspects of readability that are truly measurable,
but length of line is one. It has been established that the most frequent
destroyer of readability is excessive line length. The reason is quite obvious. Long lines of type, especially when well leaded, look graceful—just as
tall, slender models do. But the models are too skinny for anything but
modeling and long lines of type are similarly not efficient for reading.
Tests have shown many disadvantages in long lines:

a.   The eye must blink at intervals during reading. After each blink, an
     optical adjustment and a refocus of vision take place. The longer the
     line, the more frequently blinks occur within, rather than at the end of,
     lines.

b.   There is the time and visual effort lost in traveling back to the begin-
     ning of the next line.

c. When the lines are too wide, there is momentary difficulty in determining which is the next one (and sometimes the wrong one is selected). Each interruption—the blink, the trip back, and the search for the right line—causes loss of reading efficiency, or poor readability.

The optimum is about 70 characters per line in a page of average size. Fewer characters is better—down to about 50, where it becomes difficult to set justified lines without excessive hyphenation of words and irregular word spacing, both of which reduce readability.

The ideal is probably between 55 and 60 characters per line, at a length of about 4 inches (24 picas) for justified text. For unjustified lines, 45 characters is about the optimum.

Neither size of type nor length of line can be selected independently of leading. The larger the type, the more leading is needed to avoid confusion. If the space between lines is not sufficient in relation to the space between words, the horizontal movement of the eye is disturbed.

The longer the line, the more leading is needed to distinguish the lines and facilitate finding the beginning of the next one.

In general, 10-point and 11-point typefaces on line widths of up to 22 picas can do with 1 point of leading; from 22 to 25 picas with 2 points; over 25 with 3 points. Typefaces of 12 and 14 point generally need a minimum of 2 points, require 3 when set wider than 25 picas, and read better with 4 points on widths of 28 picas or over.

Small typeface sizes, such as 8 and 9 point, need proportionately more leading than the larger sizes, to compensate for their lower readability.

PAGE LAYOUT

Once the structure of the manual, as outlined in Chapter 4, has been firmly established, it is necessary to decide on the page layout best suited to the material. With the in-house preparation of text, it is not unusual to find that material is being prepared by several different people, and, unless careful attention has been paid to the instructions given to them in regard to the layout requirements, a lot of extra, unnecessary reformatting work may be required. The assignment of one writer as editor-in-charge would maintain continuity between the different members of the publications team.

The preferred procedure is for all contributors to provide their text in some word processing form and the editor-in-charge to either personally edit and format the pages or assign and monitor the work. Once a page layout has been decided on, it generally becomes the standard for the current and any subsequent manuals from the company.

TEXT LAYOUT

The arrangement of text on a page can be either single column or double column, and either fully justified (i.e., the right-hand edge of the text is vertically aligned) or ragged right (as this page is arranged). In technical manuals, more words can be fitted onto a page if the text is arranged in two columns. This is because the amount of space taken up by paragraph spacing, headings, notes, and so on is drastically reduced. To be of value, the two-column page should be fully justified with careful hyphenation.

PARAGRAPHS

The length, position, spacing and numbering of a paragraph are important to the readability of a manual text. Should the paragraph be too long, or if it follows too closely behind the previous one, there may not be sufficient divisioning of thought and the reader becomes fatigued too quickly.

Careful editing is necessary to reduce lengthy and otherwise unwieldy passages into smaller acceptable sizes with the essential unity of thought still maintained. (See Chapter 19 on editing techniques.)

Technical manuals, either on standard Imperial or ISO A4 (metric) paper, are best set in two columns with a column spacing (gutter) of 2 to 3 picas. The text is usually set fully justified and the optimum text is not less than 9/10 or not more than 10/12, the latter being the preferable setting.

Main and subparagraph headings are set flush with the left margin and the text is fully justified. Figures and tables are usually set either within the column or across the two-column width (see Figure 6-1).

When the text is prepared in single columns, there are two basic forms of paragraph layout that can be used, and for the sake of distinction we shall refer to them as either the *narrow column* or the *wide column* layout.

The narrow column method of chapter layout, shown in Figure 6-2, uses more paper than the wide column layout; however, it is more readable and

1.0 INAPALUS DOMINUS

1.1 Introvus Ipsum Eratus ad Diem

Iso dipso deum italus nibh veniam, quis comodoro ad minum introvus antiminus per lorum ipsum dolor ut nexus plebum eratus per ardua diem centum isociasus inapalus doseus erratus et non bello novumn scriptus et illustribus ipsum vel eum ututis consequat velit eum deum italus nibh veniam, quis comodoro ad minum introvus antiminus per lorum ipsum dolor ut nexus plebum eratus per ardua diem centum isociasus inapalus doseus erratus et non bello novumn scriptus et illustribus ipsum vel eum ututis consequat velit eum deum.

1.2 Ardua per Loerium

Italus nibh veniam, quis comodoro ad minum introvus antiminus per lorum ipsum dolor ut nexus plebum eratus per ardua diem centum isociasus inapalus doseus erratus et non bello novumn scriptus et illustribus ipsum vel eum ututis consequat velit eum deum italus nibh veniam, quis comodoro ad minum introvus antiminus per lorum ipsum dolor ut nexus plebum eratus per ardua diem centum isociasus inapalus doseus.

FIGURE 1-1
Mobilus Caravanus

plebum eratus per ardua diem centum isociasus inapalus doseus erratus et non bello novumn scriptus et illustribus ipsum vel eum ututis consequat velit eum deum italus nibh veniam, quis comodoro ad minum introvus.

Antiminus per lorum ipsum dolor ut nexus plebum eratus per ardua diem centum isociasus inapalus doseus erratus et non bello novumn scriptus et illustribus ipsum vel eum ututis consequat velit eum deum italus nibh veniam, quis comodoro ad minum introvus antiminus per lorum ipsum dolor ut nexus plebum eratus per ardua diem centum isociasus inapalus doseus erratus et non bello novumn scriptus et illustribus ipsum vel eum ututis consequat velit eum deum italus nibh veniam, quis comodoro ad minum introvus antiminus per lorum ipsum dolor ut nexus plebum eratus per ardua diem centum isociasus.

TABLE 1-1
Novum Scriptus

COMODORO	VEL	CENTUM
10067	90.002	10
23477	30.011	90
91003	5.000	05
22221	22.222	10

Erratus et non bello novumn scriptus et illustribus ipsum vel eum ututis consequat velit eum deum italus nibh veniam, quis comodoro ad minum introvus antiminus per lorum ipsum dolor ut nexus plebum eratus per ardua diem centum isociasus inapalus doseus erratus et non bello novumn scriptus et illustribus ipsum vel eum ututis consequat velit eum deum italus nibh veniam, quis comodoro ad minum introvus antiminus.

Per lorum ipsum dolor ut nexus plebum eratus per ardua diem centum isociasus inapalus doseus erratus et non bello novumn scriptus et illustribus

Inapalus doseus erratus et non bello novumn scriptus et illustribus ipsum vel eum ututis consequat velit eum deum italus nibh veniam, quis comodoro ad minum introvus antiminus per lorum ipsum dolor ut nexus plebum eratus per ardua diem centum isociasus inapalus doseus erratus et non bello novumn scriptus et illustribus ipsum vel eum ututis consequat velit eum deum italus nibh veniam, quis comodoro ad minum introvus antiminus per lorum ipsum.

Dolor ut nexus plebum eratus per ardua diem centum isociasus inapalus doseus erratus et non bello novumn scriptus et illustribus ipsum vel eum ututis consequat velit eum deum italus nibh veniam, quis comodoro ad minum introvus antiminus per lorum ipsum dolor ut nexus plebum eratus per ardua diem centum isociasus inapalus doseus erratus et non bello novumn scriptus etusus

FIGURE 6-1
Dual Column Setup with Table and Graphic

1.0   INTRODUCTION

This design procedure manual deals specifically with the design of central (or fixed) industrial vacuum cleaning systems found in the SEC's fossil-fired and nuclear generating stations.

The primary function of the central vacuum cleaning system is housekeeping. The removal of dust from the station floors, equipment and steel structure minimizes fire, explosion and health hazards in addition to reducing maintenance costs on equipment which may malfunction due to dust contamination.

2.0   EXTENT OF SYSTEM

The central vacuum cleaning system consists of the following basic elements:

a. Exhauster (vacuum producer).

b. Centrifugal separator and bag filter.

c. A transport piping system with inlet valves.

d. Manual cleaning tools and flexible hoses.

e. Dust unloading system.

3.0   DESIGN REQUIREMENTS

3.1   Systems Classification

Dry-type central vacuum cleaning systems may be classified into the three basic categories - light, medium and heavy-duty. The system rating is dependent upon the following criteria:

a. Quantity of dust to be removed on regular basis:

b. Characteristics of dust handled:

    (1) bulk density

    (2) particle size

    (3) abrasiveness, and

c. Vacuum required to convey the material through the system.

3.1.1   Light-duty

Light-duty central vacuum cleaning systems are typically found in large private residences, schools, hospitals, institutions, and commercial buildings.

3.1.2   Medium-duty

Medium-duty systems are found in general manufacturing facilities to remove normal floor deposits of dust and for equipment cleaning. In addition, they may be found in facilities where dust is created by the process, for example, flour and feed mills, where the dust has a light or medium density.

3.1.3   Heavy-duty

Heavy-duty systems are frequently found in foundries, ore and coal-processing plants, where large quantities of dust may be produced, probably of high-bulk density and/or abrasive characteristics.

In SEC generating stations, central vacuum cleaning systems for general cleaning of the powerhouse shall be designed for medium-duty service. Systems installed for cleaning coal- or ash-handling areas of fossil-fired generating stations shall be designed for heavy-duty service.

1-1

FIGURE 6-2
Full Page, Narrow Column Layout

provides a better aesthetic appearance. This is an important factor to take into consideration if the manuals are intended to serve as training manuals as well as to be used in the field.

In a narrow column page, each paragraph is placed flush on the text margin with the first decimal place of the paragraph number positioned over the number guideline. The headings for main and subparagraphs should stand alone and be numbered. All other paragraphs should be left blank.

Three spaces are left between the last line of text on the previous paragraph, and the following heading. Thereafter, two spaces are left to the first line of text, and this is the case for all following paragraphs not headed.

The wide column page layout may take either of two forms: the first, shown in the top part of the example in Figure 6-3, is similar in many ways to the narrow column layout. Headings are indented and numbered in the same manner, but the text is aligned with the left-hand figure of the paragraph number. This configuration ensures maximum utilization of space on a manual page.

The second style, illustrated in the lower part of Figure 6-3, is generally used when the numbering of every main paragraph is necessary (sub-subparagraphs are treated the same as for narrow column pages). This form is often found in specifications or other documents having legal aspects.

HEADINGS

The first page of any part, section, or chapter should always be a right-hand or *recto* page. The heading of the page is centralized and spaced as shown in Figure 6-3. However, in some circumstances, it may be required that the heading or titles appear on a separate *recto* page, with the text following on the left-hand or *verso* page. This does use extra paper and may be significant in a manual with many divisions.

The headings should never be higher than the top line of text in any following page.

The size and font of the typeface used for main headings should be about 2 points larger than the headings used with the main paragraphs and subparagraphs.

STATE ELECTRICITY COMMISSION

SERVICE WATER SYSTEM

PART 1

1.0 INTRODUCTION

The service water (SW) systems include all the components that are required to screen, pump, control, and convey the service water from the forebay to the end user and back to the lake.

This manual discusses the design parameters, limitations on water velocities in the various components, the computations which have to be made and the sources of information and formulas.

2.0 LEGAL REQUIREMENTS

2.1 Power Piping Code

2.1.1 All service water piping between the pumps and the equipment shall conform to the Power Piping Code B31.1. This code does not apply to systems having a design pressure of 30 psig or less, such as vented service water discharge piping.

2.2 Ministry of Consumer and Commercial Relations (MCCR)

2.2.1 For each individual power station, the flow diagrams must be submitted to the Ministry of Consumer and Commercial Relations (MCCR) for approval. The submission of flow diagrams is the responsibility of the system design engineer. The MCCR will advise if further details are required for approval.

2.2.2 The following data are to be given on the flow diagrams:

      (a) Class of piping design
      (b) Fluid handled
      (c) Design pressure
      (d) Operating pressure
      (e) Operating temperature
      (f) Test fluid
      (g) Test pressure
      (h) Minimum duration of test pressure

2.3 Hydrostatic Tests

2.3.1 Pumps taking water directly from the lake must be hydrostatically tested to 1.5 times the shut-off head of the pump (plus static head if applicable). The test is to be witnessed by the SEC inspector. The ASME and Power Piping Codes are not applicable.

2.3.2 For the high pressure pumps or any other booster pumps in the system the hydrostatic test pressure of the casing must be at least 1.2 times the sum of the shut-off heads of the primary pump plus the booster pump. This is a minimum requirement.

2.3.3 The test is to be witnessed by the SEC inspector. The ASME and Power Piping Codes are not applicable.

2.4 Certificates and Registration

2.4.1 Pressure vessels that contain gas or air are under the jurisdiction of the boiler inspection branch of the MCCR. Water hammer arrestors having an internal volume of more than 1.5 ft^2 are considered as pressure vessels under the MCCR regulations.

2.4.2 All such pressure vessels must be properly certified and registered.

FIGURE 6-3
Full Page, Wide Column Layout

NUMBERING

Paragraph Numbering

The headings of main and subparagraphs are numbered using the system of chapter number, decimal place, followed by paragraph number (followed by a further decimal place and subparagraph number if necessary); hence a reference for these may appear as 3.1 and 3.1.1, respectively.

Main paragraph headings should be fully capitalized, in boldface, with any subparagraphs and sub-subparagraph headings having boldface leading capitals only. Lead-in headings should be in bold italics. For example:

6.1 MAIN PARAGRAPH

6.1.1 SubParagraph

Lead-in Heading: These are the three basic forms of paragraph treatment.

The first paragraph of a section should be numbered 6.0 or 6, never 6. Usually the plain number (e.g., the single numeral 6) is used with a large header and separated from the text by a bold line.

Sub-subparagraphs, with or without headings, are indented and generally always referenced with a lowercase letter followed by a period (e.g., a., b., c.). They can be set in the manner of a lead-in heading but do not use bold, only italics, as illustrated below.

Any further subdivision of a paragraph should be referenced with a numeral within parentheses, for example, (1), (2), and (3). This reference is positioned directly below the left margin of the sub-subparagraph.

The sub-subparagraph is indented four spaces to the letter reference, followed by two further spaces after the period of the letter; for example,

a. *Sub-subparagraph:* Also shown is the method of adding a heading to this paragraph. It does not stand alone but forms part of the line, is italicized, and is followed by a colon.

   (1) As mentioned above, the next subdivision of the text is positioned thus. Headings, if necessary, follow the same rules as before.

By using references that alternate between numbers and letters, a reference such as 3.1.la(4) cannot be misunderstood even if the parentheses were inadvertently omitted, that is, 3.1.la4. If the sequence of subparagraph and sub-subparagraph referencing were to be reversed, deciphering the reference 3.1.lld would lead to misinterpretation.

Page Numbering

In order that adequate cross-referencing and indexing can be provided in a technical manual, good pagination (i.e., page numbering) is vital. Manuals that are complex in structure do not generally use the simple consecutive page numbering system of the textbook form.

The positioning of the page number (also called a folio) is essentially a matter of choice. The lower edge of a page is the most accepted position, especially with manuals that include some type of fold-out page. This location ensures that, no matter what form the page takes, that is, plain letter-size (8½ x 11-inch) or longer 11-inch high foldout drawing, the page number will always fall approximately in the same position throughout the manual.

The page number should be preceded by the number of the section or chapter to which the page belongs, for example, 3-6, which indicates Chapter or Section 3, page 6.

This style is the most general in use and can be applied without difficulty to most manuals: however, sometimes when a manual is made up of many parts (see Figure 4-2), there may be more than one set of sections or chapters.

This does not generally cause difficulties to the user once the manual has been bound and issued, but it can be a headache to those concerned with keeping track of pages during the preparation of the manual.

This problem can be overcome by the addition of a prefix to the front of the two-part page number; for example, if we write our 3-6 as 4/3-6, it now represents a page 6 in a Chapter (Section) 3 within a Part 4. Some manual writers prefer to add the notation in the form of wording, that is, PART 4 or even the subsystem name placed somewhere on the upper or lower edge of the page.

Figure 4-2, which shows a complex manual format, uses this system, and the numbers at the bottom of the part and section blocks together provide the reference to be used.

If this notation were arranged in the manner of 4-3-6, it could cause confusion as it is not immediately apparent which end of the notation represents the top level number. The oblique stroke isolates the top level number from the easily recognized 3-6 notation.

Figure and Table Numbering

The numbering of drawings and tables in a technical manual is essentially the same as that of page numbering: however, the reference notation does not normally extend beyond two-digit sets because the accompanying page number is sufficient to provide the location reference to the manual assembler.

The sequence of figure or table numbers would start at 1-1 and progress through to the end of the chapter or section after which the sequence would begin again as 2-1 and so on, as is done in the page numbering.

Front Matter Numbering

The pages that make up the front matter of a technical manual are usually numbered with lowercase roman numerals, that is, i, ii, iii, iv, and so on. This small section, which may include the preface, warranty, first-aid information, and all the content indexes and references is not usually prepared until after the main text has been completed; hence the use of a different page numbering system is necessary. The contents of this front part of a manual is dealt with in detail in Chapter 9.

Blank Page Numbering

The treatment afforded manual pages that are blank is not specific. In some military handbooks, the blank sheet is given its appropriate page number plus a message, such as

"THIS PAGE IS INTENTIONALLY BLANK"

This situation is sometimes encountered when a manual contains confidential or sensitive material and it is essential that every page can be accounted for.

It is sometimes unfortunate that last minute changes have to be made to a manual either just prior to printing or just after. Should drawings or text be removed, the chapter and paragraph numbering cannot be easily amended.

Several pages can be rearranged, but invariably one finishes up with enough space to fill one page. By the insertion of a numbered page with a message, such as

"THIS PAGE IS INTENTIONALLY BLANK
Paragraphs 3.2 to 3.7 inclusive have been deleted"

the existence sequence of the chapter or section has not been significantly changed.

Another method of handling blank pages in a manual, where deletions of text or other material have not occurred, is in the form of projected pagination.

In almost every instance, a blank page falls, or is made to fall, on the *verso* or left-hand page, that is, at the end of a chapter or section and on the reverse side of a large pull-out drawing. On the *recto* page preceding the blank page, the numbers of both the *recto* and the *verso* page are printed together, for example, 3-17/3-18 or as in the complex manual formats 5/3-17:5/3-18 as the case may be.

WARNINGS AND CAUTIONARY NOTICES

During the preparation of the manual text, it is necessary to consider the need for interparagraph warnings and cautionary notices (or other headed notes). These are included within the body of the text in order to bring the reader's attention to some aspect of safety or procedure.

To achieve the maximum visual effect, these notices should never be allowed to blend into the text. It is essential, in the instance of those notices that deal with the safety of personnel or possible damage to equipment, that they are immediately recognized and acted on.

Ideally, these notices should stand alone and be indented beyond the text margin with a corresponding amount of space at the opposite end of the text. NOTE: Some manuals are prepared with these important notices buried within the text where they have little impact. If this note were prepared correctly, it would appear in the following form:

NOTE
Some manuals are prepared with these important
notices buried within the text where they
have little impact.

It is quite obvious that the latter presentation of the notice is far superior and gains the full attention of the reader.

The ideal spacing for these notices is two spaces down from the preceding paragraph to the headline, which should be underlined if typed or left as is if a heavy typeface is used. Two spaces are left between the heading and the first line of text with the same spacing to the first line of text after the notice.

TITLES AND LOGOTYPES

A great number of technical manuals being produced today have the company name and address plus the logotype emblazoned on one or more edges of each page. This is often the result of letterhead stationary being used for the preparation of camera-ready copy.

While this advertising medium enhances such documents as proposals, data sheets, and sales brochures, it does tend to overpower a formal manual presentation, and its use should be carefully considered in this light.

A point worthy of note in this regard is that if a letterhead sheet is used for the preparation of text, and the logotype and/or name has been printed in color, this feature is lost if straight single-color printing of that master is performed; also, the result is often a light washed-out looking area of an otherwise well-printed page. This feature can only be used to its best advantage if the logotype and/or text is added during the printing process, or actual letterhead paper is used for printing of the manual; however, this paper is usually too light for manual pages. Both of these procedures are costly.

It is recommended that plain white paper be used for all pages in a manual and that the logotypes and so on are only presented on the cover and on the title pages of the manual (see Chapter 18 for a review of paper types).

CHAPTER 7

MANUAL WRITING STYLE

INTRODUCTION

Most manuals are separated into sections, each of which conveys a set of instructions or data for a particular activity or purpose. Each section is long enough to cover the subject and may be half a page long or even extend to 40 pages. Material is not repeated, except possibly to a minor extent and for a specific reason. Material that is logically separable into two or more sections is not combined into a single section nor is a large section broken up into smaller ones for consistency of lengths.

When a company publishes a large number of different manuals, it usually finds that most of the equipment it manufactures can easily be described within a format that changes little from one manual to another. Several advantages accrue from the use of consistent section organization, similar types of graphics in all manuals, and similar reference data in the front matter and index. Customers tend to find necessary data more easily once they get the feel of the manuals' organization. Sales personnel also enjoy the same advantage; during a sales presentation this can be invaluable both because of the availability of data and as a demonstration of the competence of the salesperson.

Another possible benefit is the effect that consistency has on the image of the company's reliability. Although not necessarily related, consistency in the presentation of data is frequently construed as resulting from firm and stable management policies and procedures.

In Chapter 3, the types of information that could comprise a comprehensive set of manuals were outlined. This set could be made up of major parts containing:

Part 1: Operation.

Part 2: Maintenance (including theory of operation).

Part 3: Parts Lists/Spares Information.

Part 4: Installation.

OPERATION

There are many types of operation manuals. The contents are not necessarily similar. The manual's basic function is to instruct how to manipulate controls such that the equipment will operate. The following basic elements would be expected in a complete operation manual:

a. Front Matter.

b. Introduction.

c. Description.

d. Operation—including operating controls and adjustments.

e. Troubleshooting.

f. Minor Servicing.

Front Matter

Front matter includes the title page, the table of contents, a list of illustrations, a list of tables, and a frontispiece. It also may include a preface and a warranty statement.

Title Page: The title page indicates clearly the subject of the manual, the equipment for which the manual is written, and other pertinent identification data. It will show the name of the equipment manufacturer. Even a single model may be changed from one order to the next and these changes may be extensive enough to require different manuals.

Sometimes the model differences are minor. Variations in the operation or maintenance can be noted in appropriate locations within the text. In commercial practice, minor differences are not usually mentioned if operation is not affected. Acceptable commercial practice is to omit some information if its inclusion could cause confusion.

Only if models change and the manual is amended accordingly should the title page indicate the serial numbers of the equipment for which the particular manual applies.

This is sometimes needed when large numbers of rapidly changing equipment types are being built. In such instances, the title page may include a date.

If, however, the equipment is likely to remain unchanged in its basic configuration for, say, five years or more, it may be advisable to omit the date on the title page. A date may not be necessary for identification, and its inclusion may lead a customer to believe that she/he has received an "old" manual, particularly if she/he has just purchased an equipment that has been represented as being the latest on the market.

Table of Contents: The table of contents lists only the major subdivisions of the manual in order of their presentation and the pages on which the subdivisions start.

List of Illustrations: This is a listing of figure titles and the pages on which they are located.

List of Tables: This lists tabulated matter and the pages on which each is located. This list of tables, however, serves an essential function in that by merely looking up a certain table, the reader can get information in an easy to-use form.

Frontispiece: A frontispiece is a picture of the equipment, or, if a process is described, a block diagram. In either case it is a graphic representation of the subject of the manual so the reader can compare the picture with the actual object. It familiarizes the reader with the appearance of the equipment before any work is performed on it. The frontispiece is usually referred to later on in the text during a discussion of the function of the equipment.

Warranty: Commercial products are usually sold with a performance or service guarantee, or warranty. If this is the case, a statement concerning the conditions of the warranty may be included in the front matter. Figure 7-2 is a sample of one style of warranty. Any warranty statement should be approved by legal counsel.

Preface: If necessary, a preface may be included to indicate how to use the manual. Some readers may not be familiar with the use of manuals. Important terminology and the organization of the material should be described. This cannot be in a separate section; it must be right at the front where the casual reader will not miss it. Such a preface may give

HRC TYPE 225E
AIRCRAFT VHF NAVIGATION
EQUIPMENT

Halesco Radio Corporation
Markham, Ontario

84120956/ZZ/WP51/BS/93
ARC225E/3
November 1994
Printed in Canada

FIGURE 7-1
Typical Manual Title Page

WARRANTY

HALESCO Radio Corporation warrants each new airborne product to be free of defects in workmanship and material for a period of twelve months from date of original installation. A defective product will be replaced or repaired (at HRC discretion) when returned to HRC, transportation pre-paid, by an HRC authorized dealer or service agency. A statement establishing the date of installation must also accompany the defective unit.

HALESCO Radio Corporation will reimburse an HRC authorized dealer or service agency for labour charges and parts replacement incurred in the repair of defective products for a period of ninety days from date of original installation. Request for payment (or credit) must be made by an authorized HRC dealer or service agency on an HRC supplied form, number 1990A (Warranty Service Report and Invoice). Such charges shall be billed at the authorized dealer or service agency normal shop labour rates.

This warranty shall not apply to any HRC product which, in the judgment of HRC, has been repaired or altered in any way so as adversely to affect its performance or reliability or has been subject to misuse, negligence or accident. This warranty is in lieu of all other guarantees or warranties expressed or implied. The obligation and responsibility of HRC for or with respect to defective equipment shall be limited to that expressly provided herein and HRC shall not be liable for consequential or other damage or expense whatsoever therefor or by reason thereof.

HRC reserves the right to make changes in design or additions to or improvements in its equipment without obligation to make such changes or to install such additions or improvements in equipment theretofore manufactured.

HRC will make available repair components when requested by the authorized HRC dealer, using Form 1990A for these requisitions.

FIGURE 7-2
Typical Warranty Statement

instructions on how to replace pages to bring the manual up to date. It may describe special graphic symbols used within the manual. It may, in fact, cover any topic that relates to the method of using not only a particular manual but also other manuals associated with this particular equipment. However, information peculiar to a particular manual is contained in the introduction.

Introduction

The introduction lists the equipment by its formal as well as informal designation. It will briefly describe the equipment and could refer to the general description or a detailed specification sheet, located elsewhere. It tells something of the purpose of the manual, its organization, and the manner in which it is to be used with the particular equipment it describes.

With reference to the latter, it may state that the manual is for field use only, for use with the equipment in the field and for checking after repair at a service depot, or for some other significant aspect of operation. It should state the approximate minimum level of training the operational personnel should possess and in this regard it may list special tools or equipment needed to maintain the equipment on an emergency basis. It is brief.

The introduction should mention some of the more significant features of which the user should be aware. Important new changes in the state of the art of the equipment should be mentioned. The user will then be more knowledgeable in his approach and will not automatically assume that they are dealing with the same sort of equipment he has in the past. This point may also be helpful to make the manual a good selling tool.

This section should not contain words that convey values or opinions; nevertheless, a certain amount of sales promotion can be applied here by careful writing. If the writer has taken pains to establish rapport with the reader by a careful introduction to the design concept, the reader will reciprocate by being more favourably disposed toward the equipment and its manufacturer, as well as toward reading and using the remainder of the manual.

At no time should the manual writer lose sight of the commercial motivations and omit sales information—particularly when that sales information may be valuable to the user of the equipment. Mention of a particular option or design feature may be construed as sales promotion but is also information valuable to the user. Hence it should be included, but without any undue emphasis.

Description

A general description of the equipment comes next. Certain key design features can be shown in tabular form and would include basic mechanical, electrical, or chemical specifications. Covered are such items as size, weight, power requirements, personnel requirements for operation, and any auxiliary equipment. This is not a statement of how the equipment functions, rather it is a statement of what the equipment is.

Operation

In operation manuals, the explanation of the equipment's operation does not need to be as detailed as would be required in service or maintenance manuals. In practice, if separate manuals are required, it is prudent to utilize some of the same information without resorting to time-consuming editing and rewriting. This is true despite the fact that the operator seldom needs to know much about the theory of the design of the equipment when compared to the person responsible for maintaining it.

For systems in which the operator controls the inputs and outputs of various discrete components ("black boxes"), the system-theory descriptions should include no more than the reasons for each black-box step. On a single product, the theory should include only the basic design philosophy and a functional block diagram. It should never show detailed schematic diagrams, nor should the explanation involve the function of minor components.

The operator must become familiar with the controls needed to make the equipment function. Some controls may be required only for maintenance, and a description of these controls should not be included. A brief statement relating to the maintenance aspects of these is all that is required. Illustrations, whether photographs or line drawings, will aid the operator in becoming familiar with the location of each control. See Figure 7-3.

Preoperational Adjustments: Once the equipment has been set up and the operator has become familiar with the function of each control, the next step is to put the equipment into operation. Before this, however, the operator must know what preliminary adjustment are necessary to enable the equipment to perform at its best. The adjustment may involve the manipulation of controls or may involve parts of the equipment that are not operating controls but that require adjustment or preparation for each particular phase of an operation.

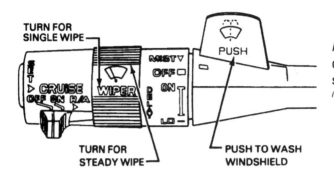

An illustration to show the use of the steering column turn signal lever in a motor vehicle.
(Courtesy of General Motors of Canada Limited)

Showing the settings of ventilation ducts in a motor vehicle.
(Courtesy of Hyundai Auto Canada Inc.)

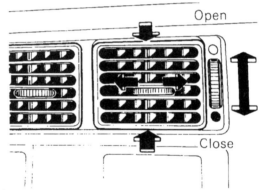

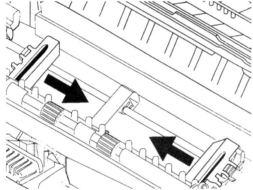

Showing an operator how to set specific paper feed path in a dot matrix printer unit.
(Courtesy of Epson America, Inc.)

FIGURE 7-3
Typical Operating Controls

The difference between operational and nonoperational controls may be too indefinite for precise definition. Normally, operating controls are those that can be adjusted while the equipment is in operation. Nonoperational controls, or adjustments, are those that must be preset before the equipment can be put into operation. Changing such controls is often the operator's job and is determined by the role the equipment is to play at that particular time.

Each mode and variation of operation should be described. Descriptions should include, where necessary, pictures or illustrations of the output of the equipment. Electronic equipment will emit a signal of a certain type and strength, whereas a mechanical device will make something or change a product. Each operation should carefully be described.

Troubleshooting

Every once in a while, a machine or piece of equipment will fail to operate or will operate poorly or improperly. It may be due to the failure of a component or subassembly. Operator error or inefficiency may be a contributing factor. It may be due to external influences such as power failure or variations in that power. At this point, it is too early to call in a technician because usually the operator has made an error or can easily correct what has gone wrong. Less frequently, the equipment will require repair.

A section with a chart showing common troubles and the remedial steps should be included. This section will not be as long or as comprehensive as the similar one in the maintenance section, however, it will contain some of the same material.

Troubleshooting for the operator will only cover those activities that are within his or her capability to perform, or those that are considered not solely technician-level activities. In many instances, operators will possess greater intelligence and capability than the level of activities allow for.

Troubleshooting instructions are written in simple language. They indicate the trouble as found, then list the probable cause/location and the remedial action. Such action may be simply *"Return unit to depot for service"* or *"Replace... with serviceable item."* More frequently, the operator will be directed to do the latter. In such cases, the operator merely calls the service department for the replacement part and the technician to replace it.

Any recalibration procedures are generally beyond the ability of most operators and invariably involve the use of sophisticated test equipment.

Minor Servicing

Such servicing can usually be adequately performed by the operations personnel. This could form part of a day-to-day start-up of the equipment and, apart from illustrations showing the orientation of the equipment, would require little explanation. Such activities may involve the replacement of filters, adjustment of video display brightness levels, or even the application of lubrication to rotating parts.

MAINTENANCE

For obvious reasons, any instruction manual must be limited in its scope. Despite the efforts of technical authors, reasonable limitations of time, money, and space tend to keep the manual to only a fraction of the length it would be if every contingency were to be covered. Many problems that will arise in the field will arise only once and cannot be anticipated or expected to reoccur.

Only the most universal conditions of operation and maintenance can be included. This is one reason for discussing the principles of operation. Should the equipment not function correctly, and the operating theory not be adequately discussed in the manual, the technician may be able to troubleshoot the problem with the information contained in this discussion.

No single maintenance manual will ever contain all the elements that are outlined here; nor is it likely that any one person will ever take the trouble to read the entire manual, except perhaps the staff engineer or the person involved with the acceptance of manuals from the manual writer. It is worthy of note that a thorough review of many good manuals will reveal that almost every manual contains at least a statement that covers each of the elements listed. A single statement may cover the requirements for two or more areas. On the other hand, a particular item, such as the setup procedures, may not be required in a particular manual simply because the subject area does not apply.

For many types of products and equipment, detailed instructions are not necessary. Considerations of prior knowledge and experience of personnel may enable some elements to be omitted. Other circumstances may mitigate against the inclusion of some elements for reasons as diverse as the following:

Legal—affecting warranties.

Fiscal—too little funds allotted for completeness.

Design—functional design being similar but precise construction different on each item within a model group.

Incomplete Data—the item may be subject to considerable redesign as a result of feedback from the field.

Here again, the importance of specific instructions bears no relationship to the length of treatment. For instance, troubleshooting information may be considered by both writer and user alike to be the most important part of the manual, yet it may occupy only 5 percent of the total space. The theory of operation may turn out to be a total waste of effort inasmuch as the details are far beyond the reader's needs.

Nevertheless, some manuals may cover this area extensively because of tariff and import difficulties of some overseas customers (comprehensive on-site repair) and because of the sales and design/description value of such material. By omitting the theory of operation, the manufacturer may place itself at a competitive disadvantage.

The listing of elements is not intended to be a complete list for any particular manual. Some equipment may require more extensive treatment in the maintenance areas. Normally separations in the manual are made between general servicing, repair without reconstruction, and the repair and overhaul (or reconstruction) function.

The following elements are among those normally included in a maintenance manual, although variations will exist from one manual to another:

a. Front matter—as in operation manuals.

b. Introduction—including statement of the purpose of the manual and manner of its use, as in operation manuals.

c. General description of the equipment—the basic configurations, as in operation manuals.

d. Modes of operation—brief statement, without specific information, on the operating techniques or procedures.

e. Operating controls and adjustment points.

f.   Theory of operation—similar to operation manuals, but if the equipment is large and complex, containing more detailed explanations of the equipment for use by skilled technical personnel.

g.   Lowest level of service required to keep the equipment functional and in operation.

h.   Periodic inspection and maintenance (service), including lubrication and the replacement of those parts with limited operating lives.

i.   Troubleshooting.

j.   Major repair.

k.   Overhaul, including procedures for the complete disassembly of the equipment down to component parts. This may cover all or only part of the equipment.

l.   Reassembly. Quite frequently, both disassembly and reassembly are combined in the overhaul section. Organization of the information is according to part or major component. The order will then be:

(1)   Disassembly

(2)   Inspection

(3)   Repair or replacement of parts

(4)   Reassembly

(5)   Adjustment

(6)   Final alignment and test

m.   Adjustment of subassemblies and components.

n.   Final adjustment of complete equipment.

o.   Checkout or test before release of equipment to operations personnel.

Any listing of elements that comprise the maintenance function must include those elements that are frequently separated, for administrative convenience, into different manuals.

Because most users of equipment want as many copies of as few different manuals as possible, most commercial manuals cover both operation and maintenance within the same volume. This saves repetition as well as costly preparation and printing. It enables both operating and maintenance personnel to use the same manual, with both groups assured of having identical information.

There can thus be no confusion as to who is responsible for a specific activity when the manufacturer of the equipment has provided the customer with all the information needed, in a suitable and accessible form, to establish one's own internal lines of control and communication.

A detailed explanation of each major element, except those stated as being similar to operating manual elements, follows:

Theory of Operation: Service personnel should understand what the equipment is supposed to do and how it does it. The manner in which the equipment operates, apart from a discussion of techniques of operation, should be described in a separate section away from theory and other details. Most of the technician's work will involve correcting malfunctions that occur during operation and cause the equipment to perform improperly.

For the technician to locate and verify the cause of the malfunction easily, he or she must know the operating modes of the equipment.

Theory-of-operation write-ups that are too complicated may induce errors in comprehension, which should and could easily be avoided. On the other hand, theory-of-operation write-ups that are too meagre may be of little value to the intended user.

The most efficient way of approaching the problem of imparting technical information to a reader is through the *graduated theory method*; that is, the whole equipment operation is first explained with a series of basic blocks, followed by a combined functional block diagram of each subassembly. Finally, a detailed full-system block diagram incorporating all important flow and control lines can be studied.

This last level is perhaps the most important for the technical reader. The explanation should impart enough knowledge to understand what happens within each subassembly without having to extend the depth of study beyond this level. Functional controls, adjustments and test points, and input/ output signals should be included with all functions and adjustments clearly explained.

Figures 7-4(a) and 7-4(b) show the typical conversion of a complex card-type subassembly schematic to a functional block diagram while retaining all essential detail. From this level on, the manual should provide, where applicable, detailed individual explanations of every subassembly. These explanations should be self-contained, that is, they do not tie in with any other associated unit except to indicate inputs and outputs.

The diagram in Figure 7-4(b) was found to be adequate for all maintenance and fault-finding activities.

Controls and Adjustment Points: As part of their knowledge of operation, service technicians must know how each control is used and each adjustment is made. Trouble often arises because of errors made by operators—not malfunctions within the equipment.

Lowest Level of Service: It has been estimated that a large percentage of all difficulties with earlier television sets (i.e., prior to the introduction of solid-state models) resulted from defective tubes. It is likely that 90 percent of the difficulties arising in most equipment lie with components/parts that are just as simple to replace as television vacuum tubes.

Experience has shown that it is probable that 90 percent of system malfunctions arise from 10 percent or fewer individual parts. In some cases, replacement can be made in the field. However, the diagnosis required to locate the trouble may involve test or bench equipment in a service center.

To keep every equipment malfunction from becoming a major project, a separate section may list the most frequent sources of trouble and the method of resolving them. It may involve an item fitted to the equipment commonly known as BITE (Built-In Test Equipment). The BITE design can range from a simple metering device to a complex microprocessor-driven test bench, packaged to fit within the equipment.

The purpose of a BITE unit is to effectively replace the multitude of individual test sets and the associated cabling that would be used to test equipment, both in the routine maintenance phase and for on-the-spot fault diagnosis. It may be an assembly of commercial test units, modified to fit within the equipment and be interconnected through internal cabling, or a complete in-house designed test set.

This schematic diagram is too complex for initial explanation – it must be simplified. The treatment effected is shown in Figure 7-4(b)

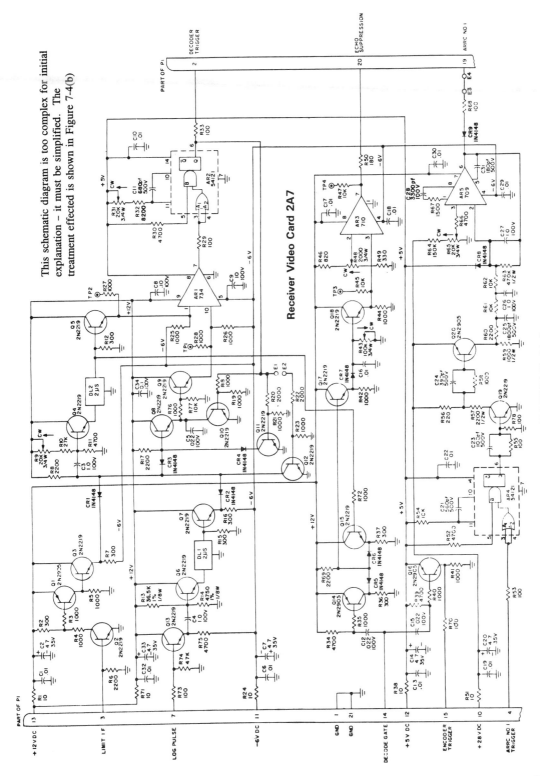

Receiver Video Card 2A7

FIGURE 7-4(a)
Block Diagram Development

63

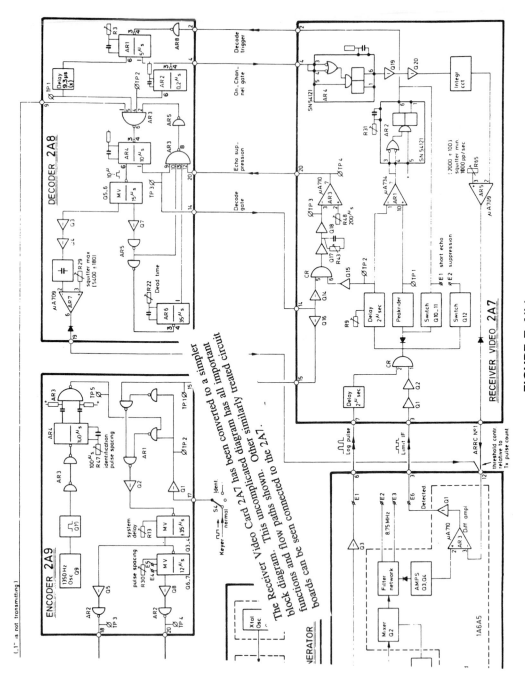

FIGURE 7-4(b)
Block Diagram Development

(.1" is not transmitting)

The Receiver Video Card 2A7 has been converted to a simpler block diagram. This uncomplicated diagram has all important circuit functions and flow paths shown. Other similarly treated circuit boards can be seen connected to the 2A7.

In the absence of a BITE unit, or even complementing it, some equipment relies on what is commonly known as a **Fault Diagnosis Chart**, whereby the most common problems are listed and a type of flowchart, originating from each listed fault, leads the service technician through a series or "tree" of sequential checks and observations.

When a fault is located by this method, the chart then either advises the replacement or adjustment procedure to be followed. In some instances, the reader is referred to a routine maintenance activity or some other section where the procedure for repair is detailed. A typical fault diagnosis chart for an automotive electric part is shown in Figure 7-5.

This does not eliminate the more complete troubleshooting procedures that may be required for even more detailed analysis of subassembly problems; however, it does eliminate much time-consuming checking and searching for solutions that can easily be attained by simple and logical fault diagnosis procedures.

The idea of this type of procedure has been the result of experience in the armed services by which they eliminated the practice of periodically replacing perfectly good parts in order to avoid future difficulties. Experience has shown that the extra maintenance created more problems than were solved. It is also now in common usage in consumer manuals.

Routine Maintenance or *Periodic Inspection and Service:* Certain products, such as automobiles, require oil changes, spark plug replacement, and other services on a scheduled time or frequency basis to ensure continuing acceptable performance. This also applies to many other products. Almost every machine with moving parts requires some form of periodic lubrication. Other products do not require lubrication at such rigidly fixed intervals.

Normal equipment usage and built-in design factors determine the frequency of periodic inspections. This type of information will be imparted to the operator so that he can recognize and inspect his machinery while it is in operation. She may also be assigned to perform the inspection and to do some of the servicing. It must be recognized that this constitutes a maintenance function.

Troubleshooting: Because so many things can go wrong in the operation of equipment, malfunctions that occur most frequently can be noted. The size of the list of troubles will depend on the specific nature of the equipment

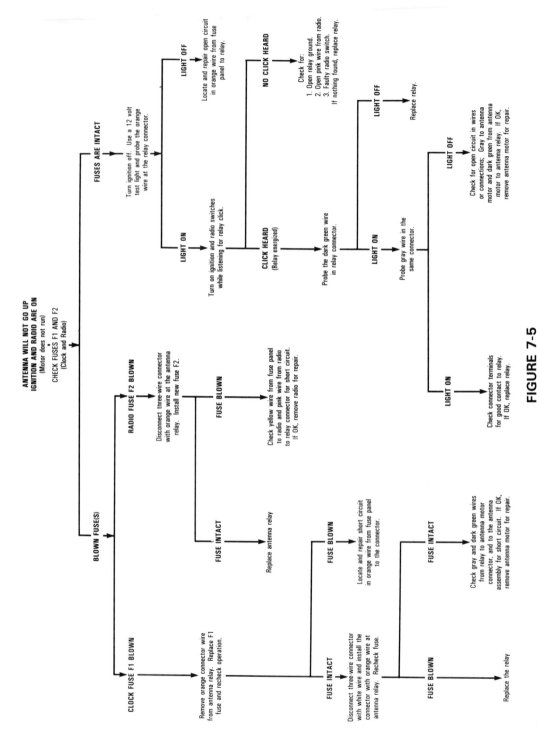

ANTENNA WILL NOT GO UP
IGNITION AND RADIO ARE ON
(Motor does not run)
CHECK FUSES F1 AND F2
(Clock and Radio)

BLOWN FUSE(S)

FUSES ARE INTACT

CLOCK FUSE F1 BLOWN

Remove orange connector wire from antenna relay. Replace F1 fuse and recheck operation.

FUSE INTACT

Disconnect three-wire connector with white wire and install the connector with orange wire at antenna relay. Recheck fuse.

FUSE INTACT

Check gray and dark green wires from relay to antenna motor connector, and to the antenna assembly for short circuit. If OK, remove antenna motor for repair.

FUSE BLOWN

Replace the relay

FUSE BLOWN

Locate and repair short circuit in orange wire from fuse panel to the connector.

RADIO FUSE F2 BLOWN

Disconnect three-wire connector with orange wire at the antenna relay. Install new fuse F2.

FUSE INTACT

Replace antenna relay

FUSE BLOWN

Check yellow wire from fuse panel to radio and pink wire from radio to relay connector for short circuit. If OK, remove radio for repair.

Turn ignition off. Use a 12 volt test light and probe the orange wire at the relay connector.

LIGHT ON

Turn on ignition and radio switches while listening for relay click.

CLICK HEARD
(Relay energized)

Probe the dark green wire in relay connector.

LIGHT ON

Probe gray wire in the same connector.

LIGHT ON

Check connector terminals for good contact to relay. If OK, replace relay.

LIGHT OFF

Check for open circuit in wires or connections; Gray to antenna motor and dark green from antenna motor to antenna relay. If OK, remove antenna motor for repair.

LIGHT OFF

Replace relay.

NO CLICK HEARD

Check for:
1. Open relay ground.
2. Open pink wire from radio.
3. Faulty radio switch.
If nothing found, replace relay.

LIGHT OFF

Locate and repair open circuit in orange wire from fuse panel to relay.

FIGURE 7-5
Fault Diagnosis Chart

and its use. Both a fault diagnosis chart and a table can be used to list the steps to take when a malfunction occurs. Figure 7-6 shows a typical troubleshooting chart arranged as follows:

a. Symptom or trouble, as indicated by poor operation, malfunction of a part, or inability to operate the equipment in all modes.

b. Cause or probable cause of difficulty.

c. Remedial action or reference to another section of the manual.

Possible troubles listed will be tabulated in a logical sequence. Any number of sequences may be used. The most probable causes are listed in the second column. The order of listing is generally in a sequence appropriate to the type of operation anticipated. The easiest solution of a probable cause is listed first, followed by solutions that are increasingly more difficult to implement.

Another logical sequence involves the listing of troubles, working down from the most frequent to the least likely source of trouble. Either way, the table is designed to ensure that the service technician will make the fewest attempts at a probable solution and make the simplest checks first.

The third column prescribes a procedure for repairing or overcoming the difficulty. Depending on the publication style, a fourth column may be added for the figure or paragraph reference for repairs if such repair procedures are carried out elsewhere in the manual.

Explanations in the troubleshooting chart are kept brief and as complete as possible without cross reference. This part of the manual is likely to be used most frequently. Even though technicians may become thoroughly familiar with the equipment, they cannot always know or anticipate every type of difficulty that other technical personnel, at other locations, have encountered. Besides the troubleshooting chart, which is the heart of this section, additional information relating to trouble diagnosis may be given in narrative form or in the form of comparison tables (e.g., current, voltage, temperature, pressure readings).

The fault diagnosis chart, shown in Figure 7-5, approaches the diagnostic routine in a different manner. Here the service technician first establishes the main identifiable symptom, then follows the diagnosis tree through a

TABLE 10-5
ARN-101D Receiver Troubleshooting Chart

SYMPTOM	PROBABLE CAUSE	ANALYSIS PROCEDURE
RECEIVER R-100		
No audio output. Other accessories connected to receiver work normally.	+100 Vdc regulated supply missing or shorted to ground.	Check Zener diodes CR6, CR7 and CR8 and capacitors A2C23 and A1C13 for shorts.
Receiver operates intermittently on one or more channels.	Misaligned crystal contacts.	Remove receiver cover. Tune to intermittent channel and manually rock the crystal drums. If fault is reproduced, check alignment of crystal contact pins.
Very weak or no audio output; IN-10 operates normally.	Improper squelch circuit connections	Check remote squelch for correct resistance values and proper connections. Check for correct pin wiring of pins E, F and S (see Figure 6-1).
	Defective audio stage	Make voltage and resistance measurements of audio stages (see Figure 5-28 for values).
First syllable(s) of voice signal is garbled or completely missing.	Diode CR2 defective producing slow squelch awakening.	Check and replace diode.
Receiver cycles continuously on some frequencies, not on others.	Contacts on crystal drum plate or control unit selector switch, or an interconnecting wire between plate and switch is shorted.	Determine which crystal drum is rotating (fractional or whole MHz). Use meter to check binary code switching sequence of the defective switching circuit.
Receiver channels correctly on less than half the frequencies selected.	Open contact on crystal drum top plate or control unit selector switch.	Determine which crystal drum (fractional or whole MHz) is at fault. Use meter to check binary code switching sequence of the defective switching circuit.
NAVIGATION OUTPUT		
IN-10 settings are not accurate.	Defective IN-10 assembly.	Substitute a known serviceable IN-10. If trouble is eliminated, replace IN-10 assembly. If trouble persists, refer to next probable cause.
	Improper phase shift in reference phase converter circuits.	Apply 40 μV 30% modulation signal to receiver. Adjust IN-10 for on-course (pointer centered) indication. Check flag current is $\geq$230 μA.
IN-10 inoperative during localizer reception. Indicator operates normally during VOR reception.	Relay K201 (in 14 Vdc model) or K202 (in 28 Vdc model).	Set control unit channel selector switches to several localizer frequencies. Check relay K201/K202 at each frequency. If relay energizes on some frequencies but not others, check the fractional MHz channel selector switch on the control unit. If relay does not operate, check if relay is serviceable; check localizer switch A1S1, and control unit switch A6S1.

FIGURE 7-6
Troubleshooting Chart

series of specific tests, the results of which lead out onto main branches. Eventually the path will lead to a smaller branch where the faulty item or subassembly can be determined. This check-and-observe method of fault tracing ensures that no routine test is omitted: one that may otherwise give a false indication as to where the fault has originated.

Major Repair: Having established the lowest level of service and having noted the troubles that are most frequently encountered, the technician must be told the procedure to follow if the repair is complex. Generally, complex repairs require quite extensive instructions; however, major repairs can be described without reference to the troubleshooting chart or to special procedures if the repair is of a general nature and possibly required frequently.

Overhaul: Overhaul is another form of major repair. It involves the rebuilding of the product with each and every part carefully checked for quality and performance. The equipment is dismantled, carefully checked at all stages of disassembly, then reassembled to an operational condition. All parts that inspection has shown to be worn, damaged, or out-of-tolerance are either repaired or replaced. Adjustments are made and calibrations performed in much the same way as was done by the factory when the equipment was initially assembled.

Reassembly: Overhaul does not necessarily require that every subassembly of the equipment be stripped down to the bare parts. Assemblies can be removed, overhauled, and reinstalled in the equipment. The process of putting components back together is called reassembly. Specific instructions may be needed for correct disassembly and reassembly of particular assemblies.

The order of reassembly may not be the reverse of the disassembly, and if not, the technician must be warned of this fact early in the procedure. The simple error of a washer being installed in the incorrect sequence has caused the crash of a commercial aircraft with resultant heavy fatalities. FAA investigation showed that the service mechanics had reassembled a particular group of parts in the precise reverse order of disassembly. They had failed to follow specific instructions, although the investigating team found that the instruction manual was not sufficiently detailed to warn of the dangers of improper reassembly.

Adjustment of Subassemblies and Components: As each subassembly is reassembled, it may require testing and adjustment. This occurs prior to reinstallation of these assemblies into the overall equipment. It is normally easier to make adjustments on smaller components individually than to make the same adjustments when the components have been installed in the larger assembly. If such adjustments are not required, it is advisable for the manual to make reference to this fact, otherwise it may be assumed that such information was accidentally omitted from the manual.

Adjustment of a Complete Equipment: Once all the subassemblies have been thoroughly checked, they must be able to operate in conjunction with one another. The overall equipment, operating as an integral unit, must be adjusted for proper operating condition. Instructions for methods of adjusting components that are interrelated must be provided.

Checkout: Now that the equipment has been completely fixed (i.e., repaired, overhauled, and adjusted), it must be tested before it can be released for field operation, where the equipment must perform as specified. If it becomes apparent that it has not been repaired properly, it must be repaired and recalibrated.

If the checkout procedure is complex, such information should be noted so as to warn the technician of possible time and test equipment requirements. Most equipment can be checked out adequately in a relatively short time with simple steps. These tests should be described in a step-by-step order for the technician to follow. If the equipment passes these key checks, it must be considered as suitable for operation.

This does not guarantee that the equipment will actually perform as specified, because seldom will a series of workshop tests ever completely simulate actual field operation. However, such tests will be identical with, or similar to, those performed in the original factory checkout.

Parts Lists: Most modern manuals have an illustrated parts breakdown (IPB) section, showing all the mechanical parts and pieces in the equipment. The most popular method of showing parts is in the exploded order of disassembly. This not only identifies the part and its position, but provides a guide to the sequence of disassembling all or a segment of equipment. The preparation of an illustrated parts lists is covered in Chapter 14.

List of Major Replacement Parts: To meet the need for identification of parts that have a limited service life and that must be replaced frequently, most commercial manuals will carry a list of recommended spare parts. This list will cover only the quantities generally needed for a single piece of equipment. If several pieces of the same type of equipment are maintained at the same location, the user may choose to order a different quantity, based on personal estimation of the potential use of the equipment.

CHAPTER 8

PREPARING A MANUAL SPECIFICATION

INTRODUCTION

When a company has the intention of producing a technical publication, either with in-house equipment (desktop publishing) or with the cooperation of a professional technical publications house, it is often difficult to establish a firm basis for the development of that manual without some form of technical manual specification being used.

With the aid of a technical manuals requirement document or a specification, it is possible to ensure that all publications, irrespective of whosoever is responsible for their preparation and publication, will be consistent in all aspects of format, layout, style, and so on, within itself and with other company manuals.

This chapter contains a sample manual specification, which, although more comprehensive than generally needed, will provide an excellent basis for the preparation of a specification tailored to the needs of any company.

The best approach in preparing a specification is to carefully examine the following requirements and extract or modify those parts that have application to the manual program, then assemble them into the new specification.

Note that, in the sample, all main paragraphs are numbered. This is an important requirement of any specification because it enables specific requirements to be clearly identified in the event of queries, objections, compliance letters, and the like. The specification should be laid out in the manner shown if only for the purposes of providing a semilegal style of document.

SPECIFICATION FOR TECHNICAL MANUALS

Table of Contents

SECTION 1—INTRODUCTION 74

1.1	PURPOSE	74
1.2	PRIORITY	74
1.3	APPLICABLE DOCUMENTS	74
1.4	OBJECTIVES	74
1.5	DEFINITIONS	75
1.6	REQUIREMENTS	76
1.7	DRAFT MANUAL	76
1.8	INTERIM MANUAL	76
1.9	FINAL MANUAL	76
1.10	CONFLICT BETWEEN DOCUMENTS	76
1.11	COPYRIGHTS AND ADVERTISING	77
1.12	PUBLICATION PLANS	77
1.13	COPY FREEZE DATE	77

SECTION 2—DRAFT MANUALS 78

2.1	REQUIREMENTS	78

SECTION 3—INTERIM MANUALS 79

3.1	REQUIREMENTS	79
3.2	PAPER STOCK	79
3.3	BINDING	79

SECTION 4—FINAL MANUALS 80

4.1	REQUIREMENTS	80
4.2	BINDERS	80
4.2.1	Outdoor Use	80
4.2.2	Indoor Use	80
4.3	PAPER STOCK	80
4.4	DIVIDERS	80
4.5	PRINTING DETAILS	81
4.5.1	External Printing	81
4.5.2	Internal Printing	81

Table of Contents (Continued)

4.6	PRINTING PROCESS	81
4.7	QUALITY	81
4.8	FORMAT	82
4.8.1	Paragraph Numbering	82
4.8.2	Headings—General	82
4.8.3	Part and Section Headings	82
4.8.4	Main Paragraph Headings	83
4.8.5	Subparagraph Headings	83
4.8.6	Sub-subparagraph Headings	83
4.8.7	Paragraph Spacing	83
4.8.8	Page Numbering	83
4.8.9	Text	84
4.9	STYLE OF WRITING	84
4.10	GRAMMATICAL MOOD	84
4.11	NOMENCLATURE	85
4.12	SAFETY WARNINGS AND INFORMATIVE NOTES	85
4.13	TABLES, CHARTS, AND FIGURES	86
4.14	FIGURE TITLES	86
4.15	REFERENCES	86
4.16	MEASUREMENTS	87
4.17	BINDING	87
4.18	MULTIVOLUME MANUALS	87
4.19	DIVIDERS	87
4.20	FRONT MATTER	88
4.20.1	Title Page	88
4.20.2	First Aid	88
4.20.3	Warranty	88
4.20.4	Table of Contents	89
4.20.5	Amendment Record	89
4.21	MANUAL STRUCTURE	89
4.21.1	General Information	90
4.21.2	Theory of Operation	90
4.21.3	Test and Alignment	90
4.21.4	Mechanical Drawings and Illustrations	91
4.21.5	Parts Lists	91
4.21.6	Installation	92
4.21.7	Repair and Overhaul	92
4.21.8	Drawings	92

APPENDIXES

A.	Publication Plan	93

SPECIFICATION FOR TECHNICAL MANUALS
SECTION 1—INTRODUCTION

PURPOSE

1.1   The purpose of this specification is to define the content, format, and quality of material comprising technical manuals produced for the ....(equipment type).... manufactured or supplied by....(Company).....

PRIORITY

1.2   This specification takes precedence over any other document relating to the content, format, and quality of technical manuals to be produced by the....(Company, hereafter referred to as XYZ).....

APPLICABLE DOCUMENTS

1.3   The following standards and specifications form part of this document:*

ANST X3.5	Flowchart Symbols and Their Usage in Information Diagrams
ANSI Y14.15	Electrical and Electronic Diagrams
ANSI Y32.2	Graphic Symbols for Electrical and Electronic Diagrams
ANSI Y32.3	Welding Symbols
ANSI 32.16	Reference Designations for Electrical and Electronic Parts
ZZZ-Z234.1-79	Metric Practice Guide
XYZ07865-01	General Specification for Technical Manual Schematic Diagrams

Application for ANSI documents should be made to the U.S. National Standards Institute Inc., 1430 Broadway, New York, NY, 1008.

Copies of the last two listed documents can be obtained by calling XYZ's Engineering Department at (411) 555-4390 (extension 422).

OBJECTIVES

1.4   These technical manuals shall be written with the view to serving the diverse needs of personnel involved with installation, instruction, operation, maintenance, and

---

* These are typical examples only; users should insert their own documents as necessary.

logistics support of the subject system. The manuals will, in many instances, serve these requirements in countries where the English language is not the primary tongue.

1.4.1 In the selection of material to be included within these manuals, it must be assumed that the latest state-of-the-art technology may not be readily available to the personnel concerned; consequently, the description of unique devices, advanced circuit techniques, or applications shall be covered in depth.

DEFINITIONS

1.5 For the purpose of this specification, the following definitions shall apply:

Customer: The purchaser of the equipment for which this manual is being prepared.

Company: The manufacturer of the equipment for which this manual is being prepared.

System: A combination of two or more pieces of equipment generally physically separated when in operation, and such other units or assemblies necessary to perform an operational function.

Equipment: Assemblies, subassemblies, and parts connected together or used in association with one another to perform an operational function.

Assembly: An assemblage of subassemblies or parts that constitute a major integral portion of the equipment such as a transmitter or receiver.

Subassembly: Two or more parts that form a portion of an assembly, replaceable as a whole, but having a part or parts that are individually replaceable, for example, printed circuit boards or terminal boards with mounted parts.

Part or Component: The smallest subdivision of a system, an item that cannot ordinarily be disassembled without destruction.

Standard Part: A standard part is one covered by any applicable MIL specification or is a commercial part manufactured as a standard catalog item, both of which are normally carried in stock by one or more manufacturers or supply establishments in major North American or Canadian cities.

Consumable Part: A part that is recognized as having a finite life of operation and that therefore must be replaced at regular intervals during the life of the equipment, for example, filters, fuses and incandescent lamps.

Special Part: A manufactured part, peculiar to the equipment or to the assemblies in the equipment, made or modified to a special design and usually with protected manufacturing rights.

REQUIREMENTS

1.6   Unless otherwise stipulated, the requirements of this specification shall be for the following:

a. A draft manual.

b. An interim manual.

c. A final manual.

DRAFT MANUAL

1.7   A draft manual shall be prepared during the development phase of the equipment and may be subject to customer review. This manual should reflect the level of technical material to be included in the final publication.

INTERIM MANUAL

1.8   Interim manuals shall be required under the following circumstances:

a. Where the final manual shall not be ready for distribution at the time of the equipment delivery. Its content level shall be determined by the customer at the time of contract negotiations.

b. When a formal training class is to be given prior to the delivery of the equipment and/or manuals. It is intended that the interim manual will afford an opportunity to assess the material contained within the draft and make corrections based on this analysis prior to the printing of the final manual.

FINAL MANUAL

1.9   The camera-ready copy of the final manual shall contain all text and illustrative material required for the manual as listed in this specification.

CONFLICT BETWEEN DOCUMENTS

1.10   Where a conflict exists between a customer's manual requirements and the contents of this specification, this document shall take precedence.

Variance from this specification is to be agreed on by both XYZ and the customer, having regard to the implications of nonstandardization within XYZ technical manuals.

COPYRIGHTS AND ADVERTISING

1.11 Copyrighted material shall not be incorporated into these manuals without the written permission of the copyright owner. Except for the identity of an equipment manufacturer, material prepared in accordance with this specification shall contain no advertising matter.

PUBLICATION PLANS

1.12 Where requested by a customer, a publication plan shall be prepared in accordance with the outline detailed in Appendix A to this specification.

COPY FREEZE DATE

1.13 The copy freeze date shall be that date set by XYZ and mutually agreed to by the customer and/or publications department concerned with the printing of the manual, after which no further changes can be made to the manual-in-preparation. Such additional material and alterations should be accumulated and presented as an amendment to the printed publication.

SPECIFICATION FOR TECHNICAL MANUALS
SECTION 2—DRAFT MANUALS

REQUIREMENTS

2.1 A draft manual shall provide the customer with a means of assessing the general technical content of a publication. It should contain narrative text, illustrations, and tabular matter essential to the clear understanding of the design, operational, and maintenance aspects of the equipment. The draft manual is not required to contain spares information or be prepared in a specific format other than that detailed below.

2.1.1 The submission requirements for a draft manual are as follows:

a. The technical content of all draft manuals must be arranged into the specified parts, sections, or chapters as detailed in the associated publication plan.

b. In order to readily locate chapters or sections within the draft manual, tabbed dividers shall be inserted between each of the above.

c. The text shall be submitted on double-spaced typewriter or dot-matrix generated pages, and, where necessary, handwritten amendments may be included but limited to no more than five changes per paragraph. Such amendments must be fully legible under any reproduction process.

Diazo or ammonia-vapor methods may be utilized for drawings supplied as part of the draft submission. To aid handling, these drawings should be reduced to 11 × 17-inch (folded), where possible.

d. The draft manual may be bound in simple binders, but where several binders must be employed to contain the material, it is essential that each cover is marked to indicate the contents.

e. Pages must be numbered either by machine or by hand and a common location must be used for all material in submission.

f. A detailed table of contents shall be supplied for the draft manual and must be located at the front of the first volume or any multivolume set.

SPECIFICATION FOR TECHNICAL MANUALS
SECTION 3—INTERIM MANUALS

REQUIREMENTS

3.1   An interim manual shall provide technical information relating to all aspects of theory, operation, installation, and support of the equipment in advance of the final manual. This publication shall be expedient to the promotion of technical understanding of a system prior to, and during, installation and initial operation; and to the support of training programs as required by the customer.

3.1.1   The contents of the interim manual shall conform to the general requirements of this specification, with the following exceptions:

   a. Pages should be prepared on a good quality printer; text, figures and tables may appear on one side only. This shall be the right-hand side of the open manual.

   b. Headings and emphasized terms normally appearing in boldface may be normal typeface and underlined for emphasis.

   c. Illustrations may be unretouched originals from a drawing department, reduced to 11-inch height by some appropriate length and printed on white paper. (Diazo or ammonia-vapor reproductions are not acceptable.)

   d. Spare part listings and associated identification drawings are not required.

PAPER STOCK

3.2   The paper requirements for the interim manual shall be $11 \times 8\frac{1}{2}$-inch white bond paper of 20 lb (75 g/m^2) or better.

BINDING

3.3   Interim manuals shall be bound in accordance with the customer's requirements. When such manuals are to be used for internal use, binding shall be in accordance with XYZ policy.

SPECIFICATION FOR TECHNICAL MANUALS
SECTION 4—FINAL MANUALS

REQUIREMENTS

4.1   This section covers the general style and format requirements for the preparation of the final manual. The following paragraphs describe the methods to be used in the preparation of the material used.

BINDERS

4.2   Binders for manuals, prepared under the terms of this specification, shall be no greater in thickness than 2-inches (50 mm) and shall be of the standard nonlocking D-ring style with three rings on 4½-inch (108 mm) centers. Page retainers shall be included.

Outdoor Use

4.2.1   Binders prepared for outdoor use shall be constructed of polyethylene material, of thickness ....[either 0.055 inch (1.40 mm) or 0.090 inch (2.30 mm) is standard] with the titles and logotypes silkscreened on the cover and spine in accordance with the layout shown in Appendix.... The binder color shall be .....

Indoor Use

4.2.2   Binders for manuals used essentially indoors or in clean environments shall be constructed of white heavy grade vinyl throughout with a clear facing on the front cover and spine for the insertion of title sheets. Front cover access for the title sheets shall be from the top side.

PAPER STOCK

4.3   Paper stock shall be 70 lb (140M/104 g/m²) offset white or better of standard 11-in. by 8½-in. letter size. Foldout pages shall be 28 lb (56M/105 g/m²) ledger paper with 50 percent rag content.

DIVIDERS

4.4   Dividers shall be placed between sections or chapters of the manual and between parts where these occur within a single cover. They shall be constructed of white 110 lb (220M) Bristol board with Mylar laminated tabs and hole reinforcements. Tab color and printing style shall be determined by XYZ.

PRINTING DETAILS

4.5   All text printing shall be in either two-column or single-column fully justified form as agreed to or specified by the customer and XYZ. The specific text requirements are specified below.

External Printing

4.5.1 All text that is prepared by an external printing house shall be electronically typeset with 10-point Times Roman. Section headings shall be 11-point Times Roman bold with paragraph headings, tabular titles, and other supplementary titles set in 10-point Times Roman bold.

Titles of illustrations and tables shall be 11-point Univers (Helvetica). For dual-column format, where table and illustrations are set within columns, they will be 10-point size.

Internal Printing

4.5.2 Text prepared within XYZ premises should to prepared to camera-ready level on white high-resolution laser printer paper, specified as long grain with 24 lb (90 g/m^2) weight. The laser printer should be capable of providing a resolution of up to 300 dots per inch or better.

The text shall be 10-point Times Roman. Section headings shall be 11-point Times Roman bold with paragraph headings, tabular titles, and other supplementary titles set in 10 point-Times Roman bold.

Titles of illustrations and tables shall be as specified for external printing in paragraph 4.5.1 above.

PRINTING PROCESS

4.6   Printing of the manual shall be by photo-offset, direct-image process (photocopier), or a similar process of comparable quality. The printing shall be on both sides of the paper.

QUALITY

4.7   The quality provisions of the manual are such that it should equal the best commercial standards for the printing process and quality of paper specified in this document.

FORMAT

4.8   The manual shall be arranged into parts, sections, and chapters in accordance with the format specified in the publication plan.

Paragraph Numbering

4.8.1   Each main paragraph and subparagraph preceded by a heading shall be numbered with the section/chapter reference and paragraph number. Paragraphs beyond three digits shall not be numbered.

Sub-subparagraphs shall be indented four spaces to the reference letter, followed by two spaces after the decimal point: for example,

a. The first sub-subparagraph shall use lowercase letters, without parentheses, and each line of text shall not extend further to the left than the alignment with the first line capital as shown here.

b. This requirement exists whether the subparagraph or paragraph preceding it is numbered or not.

(1) If the requirement arises for the introduction of sub-sub-subparagraphing, the same rules shall apply as outlined above; however, references shall be in the form of numerals within parentheses as shown.

(i) Although not recommended, further subdividing of the paragraph is offset exactly the same way as (1) above; however, the references in this paragraph shall be lowercase Roman numerals.

Headings—General

4.8.2   The various headings shall consist of all capital letters or a combination of both capital and lowercase letters to provide an indication of the divisions and subdivisions of the subject matter. Matters of equal importance shall be given headings of equal rank throughout the publication.

Part and Section Headings

4.8.3   Part and section headings shall consist of leading capital letters with a chapter/part number and a thick division line and be centered at the top of the first page of that particular part/section. The first heading shall be six line spaces from the top of the page (i.e., PART 1, SECTION 2, etc.) and the second line of the heading, generally the title, shall be nine line spaces from the top of the page.

Main Paragraph Headings

4.8.4 Main paragraph headings shall be in capital letters and stand alone, flush with the left-hand edge of the text. The heading shall be three line spaces from any previous paragraph or heading, and two line spaces to the first line of text of that paragraph.

Subparagraph Headings

4.8.5 Subparagraph headings shall be in a combination of capital and lowercase letters, and shall stand alone flush with the left-hand edge of the text. The heading shall be located three line spaces from the previous paragraph or main heading, with two line spaces to the first line of text of that paragraph.

Sub-Subparagraph Headings

4.8.6 Where headings are required on sub-subparagraphs, they are to be included as part of the first line of text. The first letter of each word, with the exception of conjunctions and/or prepositions, shall be capitalized and the heading shall be placed flush with the left-hand edge of the text of that sub-subparagraph.

Paragraph Spacing

4.8.7 Where headings do not appear, the required spacing between paragraphs shall be two line spaces. All paragraphs, with or without headings, shall start six line spaces from the top of the page.

Page Numbering

4.8.8 The pages relating to front matter shall be numbered with lowercase Roman numerals, except for the title page, which shall remain unnumbered; the text pages shall be numbered with Arabic numerals. Numbering for all front matter and text pages shall be at the bottom center of each page. In accordance with standard printing practices, blank pages shall not be numbered.

When a section/chapter ends on a right-hand page, the next section/chapter shall start on the following right-hand page, consequently, the last left-hand page will be blank.

Page numbering of foldout drawings and the like shall be on the lower right-hand corner of the sheet, below the right-hand edge of the title of the contents.

Text

4.8.9   The text, other than tabular matter and parts lists, shall be prepared in a single or double column, as determined in paragraph 4.5, single spaced with both edges justified.

The total width of the text should not exceed 6.5 inches (165 mm) with the outer edge margin not less than 0.5 inch (12.7 mm) wide.

STYLE OF WRITING

4.9   The most significant element in the preparation of a technical publication is the level of technical content. This should be represented in a language free from vague and ambiguous terminology, using the simplest words and phrases that will convey the intended meaning.

All essential information should be included, either by direct statement or by reference. For maximum clarity and usefulness, there should be consistency in terminology within the same manual or set of manuals. Also, within the confines of the material being presented, there should be a consistency of organization among all similar technical publications.

Sentences should be short and concise and the spelling of words shall be in accordance with *Webster's New World Dictionary*.

Technical words shall only be used when no other wording will convey the intended meaning. Where technical words are necessary, they shall be in accordance with the appropriate technical or industrial standard (e.g., IEEE, ASME, CEMA, and ASTM).

The first letter of words that make direct reference to main assemblies shall be capitalized, for example, Local Control Unit and Emergency Power Unit. Where reference is made in the text to actual controls, the precise name of that control and/or its marking should be in bold and fully capitalized; for example, the **OXYGEN PRESSURE** control knob is adjusted counterclockwise until....

GRAMMATICAL MOOD

4.10   Second person imperative mood should be used for all procedural text in the publication (e.g., "Install the assemble so that..."). Third person indicative text shall be used for descriptive text (e.g., "When the power fails, lamp LP7A will indicate a...).

NOMENCLATURE

4.11   Nomenclature shall be consistent throughout the publication including parts lists, maintenance sections, and other related manuals or documents. Acronyms, if used, must be written out in full the first time used and should also be explained in a glossary within the manual [e.g., "the Emergency Power Unit (EPU) shall be..."].

SAFETY WARNINGS AND INFORMATIVE NOTES

4.12   In order to draw a reader's attention to certain areas of interest or dangerous situations, the key words **WARNING, CAUTION,** and **NOTE** shall be used throughout the manual. These words shall be capitalized and printed in bold typeface. The width of this special text shall be no greater than approximately 3 inches (80 mm) in order to provide the maximum reaction to the message contained therein.

The spacing requirements for these messages are as follows. Headings shall be located three spaces down from the last line of text, with the message starting one space down from the heading. After the message there shall be three spaces to the first line of the following paragraph.

The definitions and layout of the three message types are shown below:

WARNING
An operating procedure, practice, etc., which, if not correctly followed, could result in personal injury or the loss of life.

CAUTION
An operating procedure, practice, etc., which, if not correctly observed, may result in damage to, or destruction of equipment.

NOTE
An operating procedure, condition, etc., that requires special attention.

The use of alternative wording for these headings is not recommended.

TABLES, CHARTS, AND FIGURES

4.13   Tables, charts, figures, partial schematics, and so on shall be used wherever necessary to support the text throughout the manual. Small figures and schematics may be grouped to form a single page; however, these should be located, wherever possible, so that reference can be made from the applicable text without the need to turn the page.

All tables, charts, figures and so on shall be arranged, wherever possible, on pages such that the manual does not have to be turned sideways to view the illustration or read the printing.

FIGURE TITLES

4.14   Figure titles shall indicate clearly, by a brief descriptive phrase, exactly what is portrayed on the illustration. They should indicate the function or process illustrated, the nomenclature of the equipment shown, or other pertinent and quickly understood identification.

REFERENCES

4.15   The main body of the text should only apply to the equipment for which the manual is published. Coverage of optional items, variants, earlier models, an so on should be covered by appendixes or by reference to other documents.

Tables and references to consumable materials such as fuels, lubricants, and cleaning fluids should adequately be covered by manufacturer's name and other such specifications and cross-references as necessary to permit procurement from various sources without undue difficulties in identification.

Within the manual, references to figures, tables, an so on should be clear and without chance of misinterpretation; simple references to drawings should appear as "the fuel oil piping (Figure 6-1) is..." or "and referring to Figure 10-17, it can be seen that...."

When detailing assembly and disassembly procedures where there are itemized components involved, the referencing must be such that the specific item can be pinpointed without undue difficulty; for example, "the thrust ring (item 6)..." or "the thrust ring (Figure 7-7, item 6)..." as appropriate. Where the text is concerned with one drawing over a space of several paragraphs, the following reference is appropriate: "The following items and reference numbers apply to components illustrated on Figure 3-4..." and so on.

MEASUREMENTS

4.16 All measurements within the text and on tables, drawings, and so on shall be in .... (insert system) units and in accordance with the ....(insert appropriate guide—see paragraph 1.3).

Where contractual requirements so specify, these measurements may be followed by SI (metric) equivalents, contained within parentheses as appropriate. On drawings, diagrams, and so on, a separate conversion table may be supplied to eliminate unnecessary double dimensioning on the body of the drawing.

BINDING

4.17 The contents of the publication shall be bound by either of two methods, as determined by the contractor:

a. *Three-Ring Binder:* Paper shall be drilled with three 0.38-inch (9.5 mm) holes on 4.25-inch (216 mm) centers.

b. *Cerlox Binding:* This binding shall normally be limited to manual contents of 1-inch (25.4 mm) or less thickness.

MULTIVOLUME MANUALS

4.18 Where the complexity of a system is such that the associated manual cannot be contained within one volume due to either excessive size or the dispersement of the equipment, the manual may be split into two or more volumes.

While more than one part/section may be contained in any one volume, a part/section commenced in one volume should not be continued in the next. Any divisioning should be on a part/section basis.

For major systems where subsystems may be located in different areas, separate manuals shall normally be provided.

DIVIDERS

4.19 Manuals shall be divided into parts or sections as appropriate to the format. Each divided portion of the manual shall be preceded by an indexed divider with extended tab bearing the reference applicable, for example, PART 1, SECTION 2. The tabs shall be staggered to permit ease of identification.

FRONT MATTER

4.20   The front matter of the manual shall consist of the following material and shall be presented in the following order:

> Cover/Title Page
> Artificial Respiration Methods
> Electric Shock—Rescue Methods
> Warranty Page
> Table of Contents
> List of Illustrations
> List of Tables
> Amendment Record
> List of Effective Pages
> Title Page

Title Page

4.20.1   The title page layout shall be as shown at Figure 1.* The title of the system or equipment shall be centered as shown whether there are one, two, or three lines of characters in it. If the manual covers more than one type of equipment, each type shall be given.

The printing style of the title shall be 16-point Univers (Helvetica). All subtitles shall be 10-point and the XYZ address shall be of the same style reduced to 8-point size. The XYZ wordmark and logotype shall be of the size and style as specified in.... and shall be located.....

An unperforated title page shall be provided for insertion in the sleeve of the binder.

First Aid

4.20.2   The pages covering artificial respiration and electric shock rescue shall immediately follow the title page and shall be in accordance with the latest techniques of St. John Ambulance (or similar body).

Warranty

4.20.3   The warranty or guarantee page shall reflect the latest policy of XYZ and shall be arranged as shown in Figure 2.* The example shown may not necessarily reflect the latest conditions, consequently, the wording of this statement must be cleared by XYZ prior to reproduction.

---

* These figures are not included in this sample specification. Refer to relevant sections of this handbook.

Table of Contents

4.20.4   The table of contents shall be a complete index of all parts, sections, chapters, and their respective main paragraphs. Where a manual is divided into volumes under the terms of paragraph 4.18, each volume shall contain a complete index of all volumes and their respective parts/chapters/section/ and so on, by number and title. The lists of illustrations and tables shall be prepared in a similar manner.

Amendment Record

4.20.5   An amendment is defined as an update of the information contained in the publication and shall consist of either replacement pages or individual changes to the text, parts lists, or illustrative matter. Each change to the manual shall be recorded on an Amendment Record Sheet, which shall be located within the front matter of the manual.

MANUAL STRUCTURE

4.21   The contents of a system manual shall be divided into either parts or sections, each representing a main assembly of the system. Chapter/Section 1, immediately following the front matter, shall contain general information relating to an overall view of the system. It shall not be bound by any rigid subdivisioning requirements.

Each chapter/section dealing with main assemblies and so on shall be presented in the following order:

1. General Information

2. Theory of Operation

3. Test and Alignment

4. Mechanical Drawings and Illustrations

5. Parts Lists

6. Installation

7. Repair and Overhaul

8. Drawings

General Information

4.21.1 This chapter or section of the manual should include a general description of the system or equipment as a whole and should include the basic principles of operation. One or two pages tabulating the equipment's specifications should appear here. This chapter should contain details of the intended use, operating fundamentals, and any relevant limitations. Detailed theory is not required in this chapter/section; however, it should include a system block diagram.

Theory of Operation

4.21.2 This shall include technical descriptions of all equipment supplied under this specification. The principles of operation of all assemblies and modules shall be included and range from functional block diagram level to full technical descriptions. The technical descriptions should include, where necessary, partial or abbreviated diagrams to amplify a description of an assembly or part.

Text shall include adequate cross-references to diagrams so that any description can be matched to its associated drawing(s).

Only functional block diagrams and partial diagrams shall be included in this chapter. Mechanical drawings and illustrations shall be located in Chapter/Section 4, and electrical, wiring, and schematic diagrams in Chapter/Section 8.

Operational information, that is, the procedures necessary to set up and operate the equipment under normal conditions, shall be included in this section, under the subtitle *Operating Procedures*. A full description of operating controls and any operating limitations and precautions must be included.

Minor troubleshooting procedures should be included in this section. These should include common troubles and remedial information. They are to be operator-level functions only.

Test and Alignment

4.21.3 All equipment failure maintenance, testing, and alignment procedures necessary to maintain the equipment in a fully serviceable state shall be incorporated in this section. The following areas should be included:

a. Test equipment list.

b. Preventative maintenance instructions.

c. Removal and replacement of assemblies.

d. Diagrams showing test and alignment points, component locations, exploded views, lubrication points, and so on.

e. Performance monitoring and checks.

f. Trouble analysis (fault diagnosis and troubleshooting charts).

g. Adjustment and alignment.

Mechanical Drawings and Illustrations

4.21.4   This optional section is reserved for the inclusion of mechanical drawings of the equipment and ancillaries, which may be desired but do not fit into any other specific section. Examples may be component part layout drawings or the dimensions and material of fabricated parts, necessary for the installation of the equipment, but not part of the installation instructions.

Parts Lists

4.21.5   The parts list shall consist of a breakdown of the major unit into assemblies, subassemblies, and detail parts. Items made from bulk stock, such as cut lengths of cable and commercial components, are not included. The listing shall be broken down into separate sections, each containing an illustrated group assembly parts list, a part number numerical list, and a reference designation list.

The parts list shall be presented in a table form with a descriptive column listing each assembly or subassembly and its detail parts properly indented to show their relationship to the assembly or subassembly. Attaching parts are listed in the same indented column, immediately following the parts they attach.

A reference to a figure and index number in the description column shall refer to a separate illustration in which the detail parts are shown, or to an illustration on which the part is shown as an assembled item. Parts that may be limited in use to a specific optional subassembly are to be identified by notation. Codes and part numbers of manufacturers, other than XYZ, are to be included, in parentheses, in the description column.

Except for attaching parts, the quantities listed in the *Units per Assembly* column are to be the quantities per assembly or subassembly.

Attaching parts are to be annotated with the designation (AP) immediately following the description.

A list of manufacturer's codes, as referenced in the description column, shall be provided.

Installation

4.21.6 A detailed guide to the installation of the equipment must be provided.

The section shall contain unpacking instructions and any preinstallation check lists that are necessary. Installation considerations and preinstallation preparations, such as electrical power, water supplies, cable fabrication, and air conditioning requirements must be adequately covered. Full specifications shall be provided for any consumable product that may be required to be purchased locally to aid in the installation.

Repair and Overhaul

4.21.7 This section shall be provided where the equipment is of a nature that extensive repair and/or overhaul is possible. Subjects covered may include:

> Tools required (including special-to-type)
> Dismantling of assemblies
> Cleaning of assemblies
> Inspection
> Repair and replacement
> Re-assembly and initial testing
> Final assembly
> Post-overhaul testing
> Specification tables

Drawings

4.21.8 All major equipment drawings, other than those located in the mechanical drawing section, shall be located in this section of the manual. It shall include, where applicable, schematics, wiring, piping, and other diagrams pertaining to the equipment. All diagrams will be printed on one side only, concertina-folded, and will contain a blank margin page so that the whole diagram is visible when the manual is closed. The figure number and title of the diagram must be visible when the page is folded back into the binder.

APPENDIX A
Publication Plan

This appendix would contain information known as a ***publication plan***. This is prepared by the personnel responsible for the manual preparation, and its purpose is to map out the broad outline of the manual in terms of content. The best method to achieve this is to prepare a detailed table of contents.

A PERT or CPM plan (see Figure 4-4), containing branches for the preparation of the interim or draft manual, as appropriate, should be required so that the progressive construction of the manual can be monitored.

CHAPTER 9

FRONT MATTER AND
INTRODUCTORY MATERIAL

The front matter and introductory material of a technical manual generally comprise the following titles:

a. The cover page.

b. Artificial respiration procedure.

c. Action and rescue for electric shock.

d. Safety instructions.

e. Warranty page.

f. Table of contents.

g. List of illustrations.

h. List of tables.

i. Amendment record.

j. List of effective pages.

This list is not necessarily complete nor is it necessary that all subjects be included.

THE COVER PAGE

The manual cover page appears in two places: the front cover of the manual binder and the first page inside the manual. It should contain the name and reference number of the equipment covered by the manual, the company name, and, where applicable, the issue number of the manual and the effective date of issue. A company logo should ideally be incorporated. Figure 9-1 illustrates a sample cover sheet with these features included.

ARTIFICIAL RESPIRATION

If the equipment is of such a nature that an operator or technician could possibly receive an electric shock or experience an injury from equipment, it is incumbent on the technical writer/editor to include some form of immediate action instruction for artificial respiration. The most effective method is to place a one-page pictorial reference, showing the artificial respiration technique, immediately following the cover page. Another method is to provide a plasticized card, with the technique printed on both sides, located within an inside pocket of the manual front cover.

Figure 9-2 shows a typical example of an artificial respiration page. Manual organizers can get further information and similar illustrations of procedures from branches of the American Red Cross or St. John Ambulance.

ELECTRIC SHOCK—ACTION AND RESCUE

In addition to the artificial respiration information, adequate warnings must be given regarding safety and action when working in the vicinity of live circuits, that is, around voltages of 110 Vac and higher. Personnel must be warned against the attempted rescue of an electrocuted co-worker while power is still applied to the equipment. Suitable warnings are illustrated in the following section.

SAFETY INSTRUCTIONS

General safety precautions dealing with live circuits and hazardous or dangerous materials should be highly visible to the manual user. The range of situations and materials that require safety instructions and handling are:

a.   High voltages.

b.   Compressed gases.

HALESCO
GROUND INSTRUMENTATION
RECORDER

MODEL HA-7231A

Halesco Electronics Corporation
Markham, Ontario
Canada

Issue 2   March 1994

FIGURE 9-1
Typical Cover Page

FIRST AID FOR EMERGENCIES

Priority Action Approach

In an emergency situation, do not lose your head – you must not panic!
To handle the situation properly, use the Priority Action Approach.

1. Take charge
2. Call out for help
3. Assess the hazards
4. Make the area safe

5. Identify yourself
6. Assess the casualty for life-threatening conditions and give proper first aid
7. Send for help – ambulance, police, etc.

ARTIFICIAL RESPIRATION

Artificial respiration is the technique of supplying air to the lungs of a person who is unable to breathe.

1. Gently shake and shout to assess responsiveness of the casualty.

2. Assess breathing for 3–5 seconds:
   • Look for chest movement.
   • Listen for breathing.
   • Feel for breath on your cheek.

3. Call for help to attract bystanders.

4. Open the airway:
   • Lift the chin and press back on the forehead.

5. Reassess breathing.

   If still not breathing.....

6. Pinch the nostrils and blow two slow breaths into the casualty's mouth. Take 1–1.5 seconds for each breath.

7. Assess the pulse:
   • Maintain an open airway; check for carotid (neck pulse for 5–10 seconds.

8. Send for medical help.

   If pulse is detected.....

9. Continue breathing into the casualty at the rate of one breath every 5 seconds.

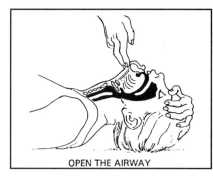

OPEN THE AIRWAY

BREATHING INTO CASUALTY

If no pulse is detected – begin Cardiopulmonary Resuscitation (CPR) if trained!

FIGURE 9-2
Artificial Respiration Page
(Courtesy of St. John Ambulance)

c. Flammable and combustible material.

d. Oxidizing materials.

e. Materials with immediate and serious toxic effects.

f. Biohazardous infectious material.

g. Corrosive material.

h. Dangerously reactive material.

i. Radioactive materials.

It is unlikely that the majority of these materials will be present in or around the equipment that the manual is being written for; however, the manual writer should be aware that circumstances may arise where these materials are present and appropriate steps should be taken to indicate the correct handling and response in the event of an accident.

In respect to the most likely situation, that is, the presence of high voltages, the following typical warning notice should appear both in the front safety sheet and within the text prior to any activity that brings personnel in contact with high voltages.

WARNING!

**HIGH VOLTAGES ARE CAPABLE
OF CAUSING DEATH**

**USE EXTREME CAUTION WHEN
SERVICING THE POWER
SUPPLY ASSEMBLY**

Figure 9-3 shows a typical safety instruction page that should be included as front matter.

SAFETY INSTRUCTIONS

The following safety instructions are of a general nature and are not related to any specific procedure and therefore do not appear elsewhere in this publication. These are recommended precautions that personnel must understand and apply during many phases of operation and maintenance.

KEEP AWAY FROM LIVE CIRCUITS

Operating personnel must at all times observe all safety regulations. Do not replace components or make adjustments inside the equipment with any high voltage turned on. Under certain conditions, dangerous voltages still exist within the equipment even with the power disconnected, due to charges retained by capacitors. To avoid casualties, always remove power then discharge and ground a circuit prior to touching it.

DO NOT WORK ON EQUIPMENT ALONE

Personnel working with, or near, high voltages should never attempt to enter an enclosure for the purpose of making adjustments to, or servicing, the equipment except in the presence of a co-worker who is capable of rendering appropriate first aid.

ARTIFICIAL RESPIRATION

Personnel working near or on high voltage equipment should be aware of the presence of the artificial respiration instructions contained in this manual, and the procedures therein.

FIGURE 9-3
Typical Safety Instruction Sheet

WARRANTY

The warranty statement, contained in the front matter section of the manual, serves to advise the buyer of the equipment of the conditions under which design changes, repair, replacement, and so on, are guaranteed. Because a warranty is effectively a legal document, its wording should be carefully phrased and if possible cleared by the company's legal counsel or lawyer. Figure 9-4 shows an example of a warranty statement for an item of aircraft equipment.

TABLE OF CONTENTS

A table of contents in a technical manual is a list of parts, chapters, sections, and paragraphs in the same order and with the exact title used in the text. There is normally only one table of contents; that is, there is no additional table of contents preceding an individual part, chapter, or section. However, each volume of a multivolume set should contain a complete table of contents covering the entire set.

The contents should be extremely legible, which requires careful spacing as well as a proper choice of typeface. Careful consideration should be taken in the decision as to the depth of the table listing. That is, if the manual uses many sub-subparagraphs, is it of any advantage to list them all? Does the subparagraph listing lead the reader to the correct area? Generally, going beyond three-digit paragraph numbering tends to make the table of contents too complex.

Parts, chapters, and sections should be spaced to show division with the leading title in capitals and in boldface. It is the custom to place the titles flush left and the page reference numbers flush right, with a row of periods, called *leaders*, connecting them. Page numbers and leaders should never be set in boldface. Figure 9-5 shows a sample first page of a table of contents.

LIST OF FIGURES (ILLUSTRATIONS)

The list of figures is handled exactly as the table of contents with respect to references, titles, and page numbers. In this instance, however, illustrations occurring in different parts, chapters, and sections are not separated. Illustrations within appendixes should be contained in the appendix table of contents at the front of that document. Figure 9-6 shows a typical list of figures occurring at the end of a table of contents. In this case, the list is

WARRANTY

HALESCO Electronics Corporation warrants each new airborne product to be free of defects in workmanship and material for a period of twelve months from date of original installation. A defective product will be replaced or repaired (at HEC discretion) when returned to HEC, transportation prepaid, by an HEC authorized dealer or service agency. A statement establishing the date of installation must also accompany the defective unit.

HALESCO Electronics Corporation will reimburse an HEC authorized dealer or service agency for labour charges and parts replacement incurred in the repair of defective products for a period of ninety days from date of original installation. Request for payment (or credit) must be made by an authorized HEC dealer or service agency on an HEC supplied form, number 1990A (Warranty Service Report and Invoice). Such charges shall be billed at the authorized dealer or service agency normal shop labour rates.

This warranty shall not apply to any HEC product which, in the judgment of HEC, has been repaired or altered in any way so as adversely to affect its performance or reliability or has been subject to misuse, negligence or accident. This warranty is in lieu of all other guarantees or warranties expressed or implied. The obligation and responsibility of HEC for or with respect to defective equipment shall be limited to that expressly provided herein and HEC shall not be liable for consequential or other damage or expense whatsoever therefor or by reason thereof.

HEC reserves the right to make changes in design or additions to or improvements in its equipment without obligation to make such changes or to install such additions or improvements in equipment theretofore manufactured.

HEC will make available repair components when requested by the authorized HEC dealer, using Form 1990A for these requisitions.

ii

FIGURE 9-4
Typical Warranty Statement

TABLE OF CONTENTS

1.0	**INTRODUCTION**	1-1
1.1	GENERAL	1-1
1.2	EXTENT OF DESIGN	1-1
1.3	STANDARD SPECIFICATIONS	1-2
1.4	SPECIFICATION DRAWINGS	1-2
2.0	**MAJOR CRANES**	2-1
2.1	GENERAL	2-1
2.2	DESIGN REQUIREMENTS	2-1
2.2.1	Classification and Duty	2-1
2.2.2	Capacity	2-2
2.2.3	Span	2-3
2.2.4	Hook Approaches	2-3
2.2.5	Lift	2-3
2.2.6	Speeds	2-4
2.2.7	Type of Hooks	2-4
2.2.8	Controls	2-6
3.0	**SMALL CRANES AND HOISTS**	3-1
3.1	GENERAL	3-1
3.2	DESIGN REQUIREMENTS	3-1
3.2.1	Classification and Duty	3-1
3.2.2	Capacity	3-2
3.2.3	Span	3-2
3.2.4	Prime Mover	3-2
3.2.5	Hook Approaches	3-3
3.2.6	Lift	3-3
3.2.7	Speeds	3-3
3.2.8	Hooks	3-3
3.2.9	Controls	3-3
3.2.10	Bridges	3-4
3.2.11	Wheels	3-5
4.0	**OVERHEAD CRANES IN NUCLEAR GENERATION STATIONS**	4-1
4.1	INTRODUCTION	4-1
4.2	CRANE APPLICATIONS	4-1
4.3	GENERAL DESIGN REQUIREMENTS—TYPES I TO III	4-2
4.3.1	Construction and Operational Periods	4-2
4.3.2	Material Properties	4-2
4.3.3	Accessway	4-2
4.3.4	Non-destructive Testing	4-3
4.3.5	Interlocks	4-3
4.4	ADDITIONAL REQUIREMENTS FOR TYPES I AND II CRANES	4-3
4.4.1	Seismic Analysis	4-3

FIGURE 9-5
Typical Table of Contents (First Page)

TABLE OF CONTENTS (Continued)

4.3	EQUIPMENT	4-13
4.3.1	Pipe Line Strainers, Y-Type	4-14
4.3.2	Forwarding Pumps	4-14
4.3.3	Circulation Heaters	4-14
4.3.4	Single-Basket Strainers	4-14
4.3.5	Flowmeter	4-15
4.3.6	Control Valves (for Displacement Pump Applications)	4-15
4.3.7	Filters	4-15
4.3.8	Relief Valves	4-15
4.3.9	Piping	4-15
4.4	CONTROL SYSTEM REQUIREMENTS	4-15
5.0	**FUEL OIL HEATING SYSTEM**	5-1
5.1	GENERAL	5-1
5.2	SELECTION OF HEATER TYPE	5-3
5.2.1	Tank Immersion-Type Heaters	5-5
5.2.2	Circulation-Type Heaters	5-9
5.2.3	Choice of Steam or Electric Heaters	5-12
5.3	HEATING REQUIREMENTS	5-14

LIST OF FIGURES

1-1	Transfer System Schematic	1-1
3-1	Forwarding System Schematic	3-1
4-1	Decision Chart	4-1
5-1	Power Required for Heating Fuel Oil	5-2
5-2	Effect of Temperature on Fuel Oil Viscosity	5-5

LIST OF TABLES

4-1	Basket Perforation and Liner Mesh for Single-Basket Strainer	4-2
4-2	Nominal Thicknesses of Shell Plates	4-2
4-3	Location of Storage Tanks Above Ground	4-6
4-4	Spacing Between Storage Tanks Above Ground	4-7
4-5	Friction Head Calculation	4-7
5-1	Pump Head Calculation	5-2
5-2	Viscosity Conversion Formulas	5-9
5-3	Viscosity Conversion Table	5-13
5-4	Specific Gravity of Petroleum Products	5-15

APPENDIXES

A	Standby Generators—Fuel Oil System	A-1
B	Power System Generators—Fuel Oil System	B-1
C	Storage Tank Plate Thicknesses	C-1

vii

FIGURE 9-6
Typical Table of Contents (Last Page)

relatively small so it is located on the last page along with the list of tables and the appendixes. Some rules for handling figure and illustration lists indicate that they should not be included if the number of entries is less than 6; however, this is entirely a matter of choice. If the list contains many entries, being on a page by itself would provide a higher degree of impact.

LIST OF TABLES

The rules for listing the tables within a manual are precisely the same as for a list of figures. Figure 9-6 shows the small list of tables contained at the end of a table of contents. The list of tables may be placed before or after the list of figures, as there is no significant ruling on this placement.

AMENDMENT RECORD

With complex items of equipment, manuals invariably undergo some form of alteration after publication, whether it be for errors in the initial preparation or for changes in an operating, maintenance, or descriptive area of the document. To control such updating, an amendment record should be provided for the customer to enter information about such changes. This record will always tell a reader what changes have been made and for what reason. A typical revision block is shown in Figure 9-7.

LIST OF EFFECTIVE PAGES

Another means of recording page changes to a manual is to provide a tabulation of each page in the manual, along with its revision status. Although widely used in military manuals, this does not have much application in civilian manuals.

Revision Number	Date of Incorporation	Incorporated By (Signature)	Details of Amendment
1	24Nov1993		Replaced pages 25 to 28.
2	03Dec1994		Value of P_{in} (para 4.23) changed from 120 mW to 300 mW.

FIGURE 9-7
Revision Block

CHAPTER 10

ILLUSTRATIONS

INTRODUCTION

The term *illustration* refers to a variety of presentation effects such as line drawings, photographs, charts, graphs, and maps. In technical manuals they are referred to, and designated as, *Figures*. *Tables*, since they are set in type rather than reproduced from artwork, are not considered illustrations.

Well-illustrated documents will be read more thoroughly by more people than the same material in text only. When illustrations are carefully chosen and executed, they support and augment the text. Frequently, the writer or technical illustrator has a rather wide choice in determining the most effective way to present such information; however, each way has its own peculiar advantages and limitations in any given application.

Illustrations are used to describe an item or idea efficiently and effectively by graphical methods. They clarify text and present phases of operational procedures difficult to describe by text alone. They call attention to details and provide graphic identification of components and tools. Illustrations can be used to depict the disassembly, reassembly, removal, and installation of equipment or parts.

A well-prepared illustration should not only present information in condensed form; it should also make comparisons easy, emphasize tendencies or trends, and make conclusions obvious to the reader.

The most common types of illustrations and their basic uses are as follows:

a. *Assembled views:* Promotional; generally depict an item of equipment in display form (Figure 10-1).

b. *Exploded views:* Used for assembly or disassembly, and for parts identification (see Chapter 14 illustrations).

c. *Operational:* Series of photographs or line drawings showing various operational characteristics of a piece of equipment (Figure 10-2).

d. *Animated:* Hand-drawn characters performing operations, in lieu of photographs or detailed line drawings (Figure 10-3).

e. *Cartoon:* Humorous characters used for impact—safety activities and the like (Figure 10-4).

f. *Procedural:* Photographic or line drawings showing how to perform specific operations (Figure 10-3).

g. *Functional:* Chart and graph style illustrations indicating data (Figures 10-13 and 10-14).

h. *Location views:* Locations of parts and equipment in assemblies (Figure 10-5).

i. *Phantom:* Parts or areas in sharp detail with surrounding areas in fadeout (Figure 10-6).

j. *Lubrication:* Equipment with emphasized locations of lubrication points or orifices (Figure 10-7).

k. *Waveforms:* Photographic or line drawings of oscilloscope waveforms (Figure 10-8).

l. *Wiring:* Functional diagrams of point-to-point wiring of equipment, often presented in color (Figure 10-9).

m. *Plumbing/piping:* Similar to above.

n. *Schematic:* Electronic diagrams using standard ANSI-Y14.15 symbols (Figure 10-10).

o. *Cutaway:* Solid drawings with pieces cut out to show inside parts or areas (Figure 10-11).

Illustrations should be as small as practical, consistent with the most effective use of space with all essential detail legible.

FIGURE 10-1
Assembly View
(Courtesy Compaq Computer Corporation)

FIGURE 10-2
Operational Illustration
(Courtesy Ontario Hydro)

Line drawings should be used wherever it is
practical. The use of a photograph over a line
drawing should be carefully considered by the
practical considerations of purpose and suitabil-
ity. Photographs, when used, should be sharp
and clear, free of shadow and cluttered sur-
roundings.

Cartoons, if used for functional purposes, should
never include copyrighted characters.

FIGURE 10-3
Animated Illustration
(Courtesy Ontario Hydro)

TYPES OF ILLUSTRATIONS

Line Drawing

Originally produced by the pen-and-ink/cut-and
paste method, this form of illustration (Figure 10-12)
is now wholly within the realm of electronics.
The illustrator has extremely powerful computer
software available, which can provide techniques previously never imag-
ined. Rough sketches or existing illustrations can now be scanned into a
computer and then redrawn to a precision determined only by the requirements of the job.

FIGURE 10-4
Cartoon-Style Illustration

Charts

Pie Charts are used primarily to show quantitative relationships between the
parts of a whole. Each part is usually expressed as a percentage of the
whole, but fractions may be used when appropriate. While pie charts are
widely used in business and departmental annual reports, their value in
making proportions readily comprehensible to the technical reader should not
be overlooked (see Figure 10-13).

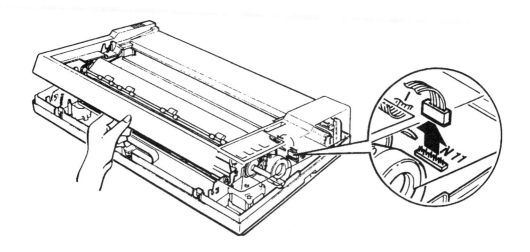

FIGURE 10-5
Location View
(Courtesy Epson America, Inc.)

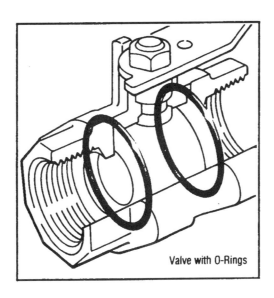

Valve with O-Rings

FIGURE 10-6
Phantom View
(Courtesy Goshen Rubber Co., Inc.)

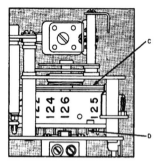

TYPICAL DRUM ASSEMBLY
SIDE VIEW

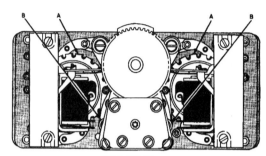

TUNER ASSEMBLY
TOP VIEW

Lubrication Point	Part to be Lubricated	Lubrication Period (Hrs)	Lubricant Type	Method of Lubrication
A	Indexing plate notches & teeth	500	Aeroshell No.7 grease	Use a sharp-pointed tool to apply a small amount to several notches and teeth.
B	Lever arm post bearing	500	Aeroshell No.7 grease	Use a sharp-pointed tool to apply a small amount into the bearing.
C	Switch plate contact surface	500	Dow-Corning 510 silicon fluid	With an artist's brush, apply small amounts to a few points on the contact surface of each switch plate.
D	Crystal pins	500	Dow-Corning 510 silicon fluid	Apply a small amount to the pins of several crystals in each drum unit, using an artist's brush.

FIGURE 10-7
Typical Lubrication Diagram and Chart

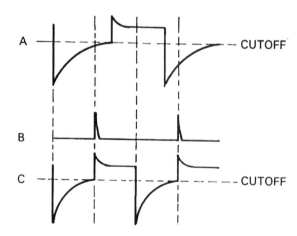

FIGURE 10-8
Typical Waveform Diagram

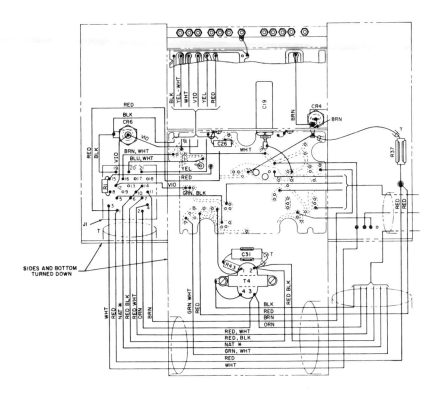

FIGURE 10-9
Typical Wiring Diagram

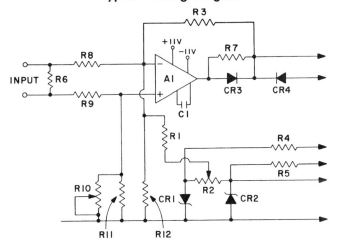

FIGURE 10-10
Typical Schematic Diagram

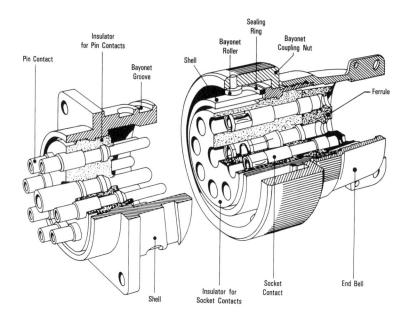

FIGURE 10-11
Typical Cutaway Diagram
(*Courtesy of ITT Cannon*)

Bar Charts present information about discontinuous phenomena or discrete quantities in a way that simplifies comparison. Bar charts may be oriented either vertically or horizontally, depending on space limitations. When space is not a limiting factor, the dependent variable should be plotted vertically, since height is conventionally associated with magnitude. By using different colors, or different black-and-white patterns, increments may be shown on bar charts, thus combining some of the advantages of pie charts with those of bars alone (see Figure 10-14).

Pictorial Charts serve about the same purpose that bar charts do. The technically trained reader, being familiar with bar charts and curves, can interpret them without difficulty; however, when technical information is graphically portrayed for a general or lay audience, pictorial charts may be more easily understood.

Two kinds of pictorial charts are in general use: one consists of two or more scaled enlargements of a stylized or generalized figure, each one of which is proportional to the value it represents; the other consists of a series of stylized or generalized figures, each of which represents a certain unit value (see Figure 10-15).

The scaled figures are useful in illustrating difference in area or volume, while the series of identical figures can be used to indicate numbers of units.

Graphs

Graphs are widely used in technical exposition because they are capable of showing the changing relationships between continuously varying phenomena.

However, their very usefulness often obscures the fact that they cannot be applied universally. Graphs should **never** be used to show discontinuous phenomena or discrete quantities.

Captions and Labels

Captions on charts and graphs must be complete enough to avoid ambiguity. Many technical writers and illustrators, in an attempt to express such information as briefly as possible, omit some pertinent information needed for complete understanding.

Captions and labels should include both the name of the variable and the unit of measure in which it is expressed. The unit of measurement may be separated from the name by a comma or dash, or by enclosing it in parentheses, for example:

Critical Pressure, psig

Speed—fps

Temperature (°F)

The practice of labeling variables with multiples or fractions of units of measure is acceptable if it is done in such a way that no ambiguity is created.

Ambiguous	*Clear*
Torque—ft-lb × 100	Torque—hundreds of ft-lb
Pressure (psi/1000)	Pressure (thousandths psi)

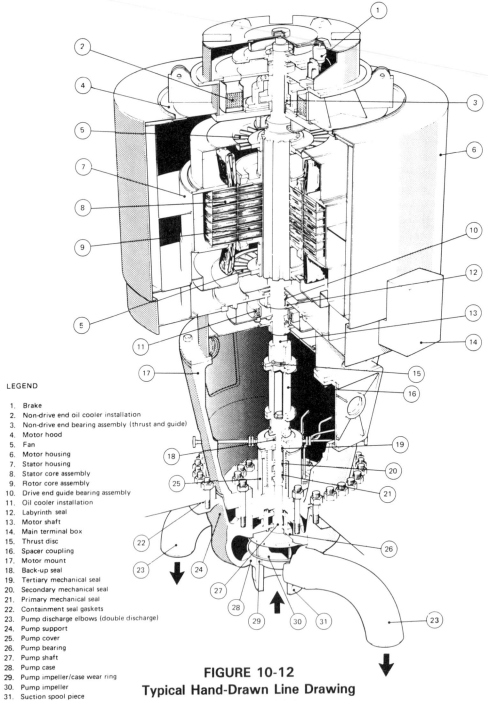

LEGEND

1. Brake
2. Non-drive end oil cooler installation
3. Non-drive end bearing assembly (thrust and guide)
4. Motor hood
5. Fan
6. Motor housing
7. Stator housing
8. Stator core assembly
9. Rotor core assembly
10. Drive end guide bearing assembly
11. Oil cooler installation
12. Labyrinth seal
13. Motor shaft
14. Main terminal box
15. Thrust disc
16. Spacer coupling
17. Motor mount
18. Back-up seal
19. Tertiary mechanical seal
20. Secondary mechanical seal
21. Primary mechanical seal
22. Containment seal gaskets
23. Pump discharge elbows (double discharge)
24. Pump support
25. Pump cover
26. Pump bearing
27. Pump shaft
28. Pump case
29. Pump impeller/case wear ring
30. Pump impeller
31. Suction spool piece

FIGURE 10-12
Typical Hand-Drawn Line Drawing
(Courtesy Ontario Hydro)

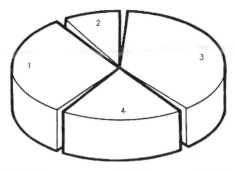

1	INDUS Inc.	31.8%	$1,278,000
2	Universal	9.5%	381,000
3	BHPL Inc.	40.0%	1,607,000
4	Martin Ltd.	18.8%	755,000

Distribution of Contracts

FIGURE 10-13
Typical Pie Chart Illustration

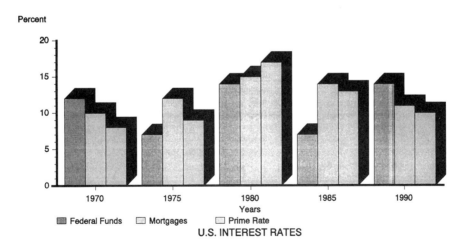

FIGURE 10-14
Typical Bar Chart Illustration

The ambiguity of the labels or captions in the left-hand column results from the reader's wondering whether the indicated multiplication or division *has been* performed by the writer or *is to be* performed by the reader. While a reader who knows enough about the subject may be able to deduce the proper value of the term, no conscientious writer or illustrator risks being misunderstood in this way.

If, when constructing charts or tables, it is found that a descriptive term or a constant decimal multiple (zeros) is repeated throughout a column or beside a coordinate, the term should be included in the heading or label to avoid useless repetition.

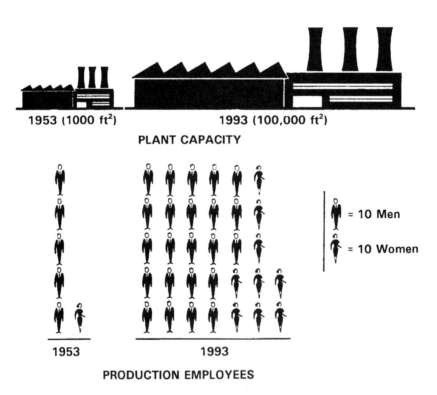

FIGURE 10-15
Typical Pictorial Chart

Choosing Scales and Grids

The proper choice of scales for graphs is a matter of judgment sometimes restricted by the amount of space available. When space limitations prevent the use of arithmetical scales for curves, the information can often be compressed by plotting it on semilog, log–log, or other special grids to bring it into usable proportions.

When the "non-square" grids are used, particular pains must be taken to call the reader's attention to that fact; the curve itself would be quite misleading if the reader were to view it as a line illustrating the relationship between simple, linear coordinates. Even when using conventional decimal-grid paper, it is necessary to use good judgment in choosing scale divisions.

Scale values for the main divisions of decimal graph paper should be limited to decimal multiples or decimal fractions of the units 1, 2, or 5; that is, each main division should be 0.0001, 0.01, 1, or 1000; 0.02, 2, or 2,000,000; 0.5, 50, or 50,000; and so on.

TIPS FOR CREATING AND USING ILLUSTRATIONS

Each type of illustration has its unique strengths and weaknesses. The guidelines presented here apply to most visual material used to supplement the technical information in the documentation being prepared. Following these tips should help create and present visual material to good effect:

a. Keep the information as brief and simple as possible.

b. Try to present only one type of information in each illustration.

c. Label or caption each illustration clearly, using the term FIGURE, followed by the reference number, to designate each one.

d. A short concise title is preferred to a wordy multiple line of text.

e. Include a key that identifies all symbols, when necessary.

f. Specify the proportions used or include a scale of relative distances, when appropriate.

g. Make the lettering horizontal for easy reading whenever possible.

h. Keep terminology consistent. Do not refer to something as a "proportion" in the text and as a "percentage" in the illustration.

i. Allow enough white space around and within the illustration for easy viewing.

j. Position the illustration as close as possible to the text that refers to it; however, an illustration should *never* appear ahead of the first text reference to it.

k. Be certain that the significance of each illustration is explained within the text.

l. Figure numbers should always be used, numbered consecutively throughout a document. The exception to this rule is in those manuals where a chapter or section stands alone and all numbering carries the prefix of the chapter number (e.g., Figure 4-3, Figure 10-5), or in appendixes, where letter prefixes should be used (e.g., Figure A1).

m. When illustrations or tables are used in a manual, list them, with their figure/table numbers and associated page numbers, under a separate heading ("List of Figures" then "List of Tables") following the text references in the **Table of Contents**.

Presented with clarity and consistency, illustrations can help a reader focus on key portions of a document. Be aware, though, that even the best illustrations only supplement the text. The text must carry the major burden of providing the context for the illustration and explaining its significance.

CHAPTER 11

TABLE PREPARATION

INTRODUCTION

A well-prepared table can present data in a more concise form than text can and is more accurate than a graphic presentation. The table can show large numbers of specific, related statistics in a small space and can provide numerous facts that cannot be conveyed in a graphical presentation. A table makes comparisons between figures easy because of the arrangement of the figures into rows and columns, although overall trends about the information are more easily seen in charts and graphs.

GUIDELINES FOR CREATING AND USING TABLES

Table Number: All tables presented in a technical document should be assigned a specific number and title, which is to be centered *above* the table (different from an illustration, which has the figure number and title *below* it). The numbers are usually Arabic, and they should be assigned sequentially to the tables throughout the text.

Tables should always be referred to in the text by table number rather than by position ("Table 4-3" instead of "the following table"). Generally, if there are more than six tables in a manual, they are listed under a separate heading ("List of Tables") with their page numbers, immediately after the List of Figures in the **Table of Contents** (see Figure 9-6). If the list is long, arranging them on a separate page would be preferable.

Typefaces: The ideal typeface for a table is 1 to 1.5 points smaller than the surrounding text when the table is combined with text. Use a *non serif* typeface such as Helvetica to improve the clarity of the small typeface.

Title: The title, which is placed just above the table, should describe concisely what the table represents. Avoid lengthy descriptions and, if necessary, extend the title details to a footnote with the use of an asterisk.

Boxhead: The boxhead carries the column headings in bold text. These should be kept concise but descriptive. Units of measurement, where necessary, should be specified either as part of the heading or enclosed in parentheses beneath the heading. With modern desktop publishing software, the box can be lightly shaded for emphasis.

Stub: The left-hand vertical column of a table. It lists the items about which information is given in the body of the table.

Body: The body comprises the data below the boxhead and to the right of the stub. Within the body, columns should be arranged so that the terms to be compared appear in adjacent rows and columns. Where no information exists for a specific item, substitute a centred em-dash (—) to acknowledge the gap.

The same unit of measure must be used throughout each column. The reason for this is obvious; the reader cannot make accurate comparisons unless data are stated in the same terms.

Rules: These are the lines that separate the table into its various parts. Horizontal lines are placed below the title, below the body of table, and between the column headings and the body of the table. Headers for specific blocks of data should be in bold capitals. The columns within the table should be separated by vertical lines to aid clarity.

Footnotes: Footnotes are used for explanations of individual items in the table. Use superscript numbers where practical, but whenever a possibility exists that the superscripted numbers could be mistaken for numerical data, use symbols (*, #) or lowercase letters.

Source Line: The source line, which identifies where the data were obtained, appears below any footnotes (when a source line is appropriate).

Continuing Tables: When a table must be divided so that it can be continued on another page, repeat the boxhead and give the table number at the head of each new page with a "continued" label, for example,

TABLE 7-3 (Continued)

Figure 11-1 is an illustration of the form and component parts of a typical table. Figure 11-2 is an example of a relatively complex table with all the major components included.

TABLE NUMBER
(Subtitle in Capitals and Lowercase Letters)

Stub Heading	Column Caption	Column Caption	
		Subcaption	Subcaption
Line Heading Subheading Subheading **Line Heading** Subheading **Totals**	Tabulated Data[1]		

1. Table footnotes and other comments.

FIGURE 11-1
Nomenclature for the Parts of a Table

Informal Tables

Informal tables provide a useful form of tabulation midway between the text statement and the formal table. An informal table should be used whenever the tabulation is very simple, consisting of not more than a stub and one column.

Tabulations of more than one column should be made into formal tables.

The informal table forms a part of the sentence structure; therefore it is incorrect to insert such a table after a sentence completed with a period.

Because it is a part of the sentence structure, an informal table is not numbered and is not given a title; the information that is normally given in the title is given in the text material introducing the table.

Example: The approximate values of surface tension, based on measurements at ordinary room temperatures, for certain selected liquids in contact with air are as follows:

	Surface Tension
Liquid	**(N/m)**
Benzene	29
Glycerine	63
Mercury	460
Water	75

Note that the stub heading and the column caption (including the unit of measure in parentheses, if it is not given in the introductory sentence) are either in bold and/or italicized. No other ruled lines are used in informal tables, although leaders (dotted lines) may be used to aid the eye of the reader. All rules governing centering and aligning are to be observed.

Tips for Creating Tables

The guidelines presented here apply to most visual material used to supplement the technical information in any tables presented within the body of text.

a. Keep the data as brief and simple as possible.

b. Try to present only one type of data in each table.

c. Label each table clearly. A short concise title is preferred to lengthy multiple lines of text.

d. Specify the proportions used.

e. Keep terminology consistent. Do not refer to something as a "cask" in the text and as a "container" in the table.

f. Position the table as close as possible to the text that makes reference to it; however, a table should *never* be located before the first text reference to it.

g. Be certain that the significance of each table is clear from the text.

h. Table numbers should always be used, numbered consecutively throughout the document. The exceptions to this rule are informal tables and tables in manuals where a chapter or section stands alone or where the table is within an appendix.

i. If tables appear in a manual, list them, together with the table reference number and page number, under a separate heading ("List of Tables") following the List of Figures in the **Table of Contents**.

TABLE 12-3
Technical Specification Data Sheet

Description					Williss Co. Specification	Company's Specification
METERS:						
	Digital	Analog	Local	Remote		
1 Input ac voltmeter						
2 Input ac ammeter						
3 Output ac voltmeter						
4 Output ac ammeter						
5 Battery dis/charge ammeter						
DIAGNOSTIC FEATURES:						
1 Status indicating lamps						
2 Microcomputer system						
3 Printer						
BATTERY:						
1 AUTO/MANUAL control for battery (YES/NO)						
2 Control facilities per B8.1.1 of technical specification (YES/NO)						
3 Type of battery to be installed (State)						
4 Number of cells						
5 Battery rating					AH	AH
6 Battery manufacturer's specification sheets for battery (YES/NO)						
7 Equalize voltage per cell					V	V
8 Float voltage per cell					V	V
9 End voltage per cell					V	V
ADDITIONAL DATA:						
1 Enclosure type						
2 Input power cable termination lug size						
3 Output power cable termination size						
4 Cable entry _____top _____bottom						
5 Audible alarm level					dBA	DBA

NOTES:
> Manufacturer's name:
> Model number:
> Type:
> Sheet 1 of 1 (Set No._____)

POWER RECTIFIERS FOR CLASS II SYSTEMS

FIGURE 11-2
Complex Table Form

CHAPTER 12

OPERATION

INTRODUCTION

The role of the operation section of a technical manual is essentially to instruct personnel how to operate the equipment about which the manual is written. It is not intended to instruct technical personnel how to service the equipment, although it may contain cèrtain procedures that could also be found in the maintenance section. These would be considered operator functions.

The complexity and manner in which the equipment is operated may justify the preparation and use of a separate operating manual.

CONTENT

If the operation section of the manual is well-written and comprehensive, the user will have no trouble understanding how to utilize the equipment. It follows that a well-informed operator is less likely to make mistakes in using the equipment, which in turn would reduce the need to call for service or repair.

Using examples of motor vehicles, a simple comparison would be the routine operations that a car owner regularly performs, for example, checking coolant, transmission oil, engine oil, and battery condition. The car owner is essentially the user but does perform some minor technical functions to generally maintain the equipment (vehicle) in an operating condition. This is generally the role performed by an operator of any equipment.

Using the example of the family motor vehicle as our item of equipment, the owner's manual is an ideal illustration of what an operator's manual would contain and how it should be used. Although a car owner's manual is almost always a completely separate document, the format would be quite suitable for inclusion as part of a complete manual.

The average motor vehicle owner's manual may contain sections such as:

a. *Introduction*: A brief page or two that could contain information relating to other manuals that are available in different languages such as Spanish or French. It may also indicate the various models to which the manual applies.

b. *Instruments and Controls*: This section would detail all instruments and controls of the vehicle. This may include radio/CD/stereo systems, fitted and optional instrument panels and/or controls (e.g., digital or analog types), and heating and air-conditioning controls.

c. *Before Driving the Vehicle*: Explaining safety features such as seating adjustments, safety belts, airbags, entry/locking conditions, alarm (theft deterrent) systems, fitted and/or optional items, and other equipment/ systems not directly related to operating the vehicle.

d. *Starting and Operating*: Details such things as "running in" procedures, fuel types, instruments and controls (usually directly related to driving), visual checks and basic maintenance, trailer-towing methods and limitations, using cruise control, and four-wheel-drive systems.

e. *In Case of Emergency*: May include how to activate the hazard warning equipment, jump starting method and precautions, what to do in case of engine overheating, towing precautions, jacking the car, and changing tires.

f. *Appearance Care*: Methods and precautions needed when using various products for cleaning both the interior and exterior of the vehicle.

g. *Emission Controls*: Because of the significance of emission controls built into motor vehicles, this section may be included as a separate section. It would detail information and precautions relative to the apparatus within the vehicle such as the catalytic converter and PCV valves.

h. *Service and Maintenance*: Would detail all routine maintenance that should be carried out by the owner or a service station, on a regular basis (tire pressures, oil levels, tire rotation, etc.). This section should include information relating to the location and replacement data for all internal/external lighting equipment (bulbs).

This section would also include reference to the routine service maintenance schedules by the dealer (inspections and service at predetermined periods, which would be issued as a separate document).

i. *Specifications*: Lists the location of vehicle identification data and basic specifications (capacities, belt tensions, fuse ratings, etc.).

j. *Service Station Information*: A brief synopsis of how to locate fillers for replacement of engine oil, coolant, washer fluid, and hood releases. This small section, located at the very rear of the handbook, provides a quick and basic information page to cope with a service station stop.

From this outline, it can be seen that the level of detail the vehicle owner gets in the owner's manual is quite extensive. Vehicle owner manuals are usually extensively illustrated to reinforce the descriptive text.

In a further illustration, for equipment destined to be installed in an aircraft, for example, a navigating instrument, the operating manual/chapter may contain:

a. Sections pertaining to operating limitations and precautions. Within this section could be subsections detailing certain atmospheric conditions/ characteristics that could be encountered during operation.

b. Tables containing data necessary for determining appropriate operating procedures under varying conditions.

c. Photographs or detailed line drawings of the actual instruments with locations of all controls clearly marked (see Figure 12-1).

d. Detailed instructions as how to perform certain operations, for example, different ways to use the equipment to navigate from one position to another.

e. Full color illustration of maps with navigated courses shown plotted with pictures of the appropriate instrument readings for each condition.

When preparing an operating manual, the writer must consider and/or investigate all activities and functions for which the manufacturer considers the operator should be responsible for. Sections should include all modes and methods of operation, test or checking functions to verify the equipment readiness, plus the results to be expected from each type of operation.

A summary of controls and indicators, prepared in a table format, is usually very helpful. If a piece of equipment has a vast number of controls and indicating devices, separate tables with accompanying drawings or photographs will help the operator's training. Care must be exercised not to infringe on any maintenance activity where conflict may occur.

Figure 3-1 C-81A Control Unit, Operating Controls

Figure 3-2 IN-10 Course Indicator
Operating Controls

Table 3-2 Operating Controls

CONTROL	FUNCTION
VOL-OFF	Combination volume control and on-off switch. Controls application of power to the Type 15F. Clockwise rotation turns the power ON. Further clockwise rotation increases the audio level of the R-34A receiver.
SQUELCH	Controls the receiver squelch circuit. Clockwise rotation silences the R-34A receiver.
MHz Channel Selector Switch (No panel marking)	Selects the MHz channel frequencies, in 1 MHz steps, between 108 MHz and 126 MHz.
Fractional MHz Channel Selector Switch (No panel marking)	Selects fractional MHz channel frequencies in 0.1 MHz steps, between 0.0 MHz and 0.9 MHz.

FIGURE 12-1
Typical Operating Controls

CHAPTER 13

MAINTENANCE AND REPAIR INSTRUCTIONS

INTRODUCTION

The maintenance and repair section, containing maintenance instructions, of the equipment manual is possibly the most important of all. It permits the owner/operator of the equipment to maintain it in a fully serviceable state for most of the life of the equipment. It also acts, to some degree, as a training manual for the service personnel because it permits technicians to gain an intimate knowledge of the workings of the equipment to the extent that downtime becomes less as expertise improves.

This section differs from the operational side of the manual inasmuch as each part of the equipment is dealt with in detail. The operator is told how to adjust a control and what result can be expected from the action. The service technician on the other hand, is told the same plus how the action is derived through the mechanical or electronic workings of the equipment.

The audience for the maintenance and repair section is vastly different from the sales and/or operational staff. The personnel using this section will almost always be technically trained and very familiar with the product. These people will be professional service technicians already trained in the equipment but, in some instances, they may also be technically knowledge-able amateurs, forced by circumstances to accept the role of maintenance service technicians.

The explanations of the different parts of the equipment and the maintenance carried out on them have to be quite detailed in nature; however, the manner in which this technical information is imparted is very important. A complex, difficult-to-understand technical description is virtually useless if the technician is forced to ponder what action is meant to be taken or what a particular adjustment setting is to be used.

The writer must never assume that technicians will have prior knowledge of anything that they are expected to repair or overhaul. For example, while a milliammeter is a common electronic component and a high percentage of technicians can successfully test one for coil resistance or full scale deflection current, many will not and could easily destroy the meter if the complete testing procedure were not fully detailed. The following subjects are generally found in this section of a manual.

EQUIPMENT CONFIGURATIONS

Often equipment is manufactured in several different models and configurations, and unless the company produces a manual for each specific one, the routine of determining where each model's set of instructions is located in the manual is time consuming. If all differing pieces of equipment utilize identical subassemblies, the problem is nonexistent.

Several methods can be used to distinguish different models or optional assemblies:

a. Where accessory type equipment is fitted, confine all information in the form of a mini-manual and locate it as either an appendix or an addendum (see Figure 4-2). In this case, some means of attesting to the fact that such items are fitted to the equipment must be used. This can be either a list, at the front of the manual, which details all assemblies, subassemblies, and accessories fitted to this particular equipment or installation, or a comment in the maintenance section introduction indicating that certain items, if fitted, are fully documented in a certain appendix or in an addendum.

b. Another way of differentiating pieces of equipment or assemblies is to use designators within the text headers, for example,

11.2 Adjustment of Relay A14K2 (Not applicable to Model 14M)

or

10.6 Receiver Phase Shift (Models Z26B and Z27C only)

The fault of this method is that it can become unwieldy if there are many models, each with different procedures.

It would be better to produce subsections with relatively common procedures and then designate them as sets or series. For example, it could be set up so that there would be a separate maintenance instruction set for:

Series A (Models A, B, C, and D)

Series B (Models K, P, and R)

SAFETY INSTRUCTIONS

The rise in the consumer movement and the increase in product liability suits in recent years have forced manufacturers to pay more attention to safety warnings that ensure the safe operation of their equipment. Given that a manual may be used as evidence in litigation, it is in the interests of the manufacturer to ensure that adequate warnings are given to both the end user and the service technician when dealing with a product.

Not only is it essential that the manufacturer make the equipment safe to operate, but it has to be relatively safe to service. If the service technician is in any way at risk during the course of a maintenance operation, it is the duty of the manual writer to ensure that sufficient warnings are given and such information is located in the correct place.

The service technician must be fully aware of any hazardous situation before he or she takes any action with the equipment. The cooperation of the company's safety officer and legal counsel are generally necessary in setting out safety measures.

There are many ways in which an equipment may pose a hazard to a technician. For example:

a. Inadequate warning of all risks or hazards that may be present in the equipment. For example, a notice that safety belts and hard hats must be used at all times when working on certain parts of the equipment, such as a microwave tower.

b. Failure to warn the service technician adequately of such risks and hazards. A brief outline of specific risks at the beginning of the manual is not sufficient. A notice should be located at any point in the maintenance instructions where the following action puts the service technician at risk.

A warning, located just before a maintenance activity that could put a service technician at risk, is the most suitable way of ensuring worker safety.

For example,

5.12 Replacement of Power Supply Capacitors

If one or more of the power supply capacitors is to be changed, first switch off main power to the cabinet, then proceed as follows:

DANGER!

The voltage stored in these capacitors is lethal!

After removing the safety cover plate, the grounding switch must be pressed *at least three times*. Use the *voltage indicator wand* on each of the positive (+) terminals to ensure that all voltage has been removed before proceeding.

---

    a. Remove the *Plexiglas* safety cover, then carry out safety measures to drain residual voltage from the power unit capacitor bank.

MODES OF OPERATION

The various modes of the equipment operation may be pertinent to the maintenance situation. Although it is assumed that the maintenance technician is fully aware and trained in the operation of the equipment, it is often necessary to provide a brief outline of the operational modes.

For example, a broadcast transmitter may have three modes of operation: Full Power, Standby, and Maintenance. All modes of operation should briefly be outlined; however, this latter mode may be such that the technician is required to place the equipment in a condition such that he or she is able to run it up to full power for testing purposes without transmitting a signal to the air. This mode may be engaged automatically, or it may require the technician to proceed through certain steps to reach the condition.

THEORY OF OPERATION

In Chapter 7, in the principles of operation discussion, it was indicated that the best method of outlining the theory of operation of the equipment was to use the graduated theory method. Using this method, the reader was provided with theory that advanced from simple block diagrams to detailed functional block diagram level (Figure 7-4). The text did not go any further than this because an explanation of the detailed workings of subassemblies, PCB cards, and so on were not necessary at this stage.

In this chapter, because the technician is concerned with both routine maintenance operations and the repair of the equipment, it is necessary to provide fully detailed explanations of the workings of every repairable subassembly or module. Armed with the preliminary knowledge of the system, and having this detailed theory of operation, along with the accompanying schematic diagrams, a competent technician should be able to fully repair any faulty part (that is deemed, by design, as repairable).

TEST EQUIPMENT AND TOOLS

Generally, with complex equipment, it is necessary to provide a table or list of the test equipment that is needed for all expected maintenance or repair. The name of the item, a manufacturer's designation, and the characteristics of the test equipment should be provided. The list should be comprehensive enough for the customer to provide an alternative piece of test equipment if necessary.

An accompanying diagram is often needed to provide the necessary layout of the test setup (Figure 13-1).

Occasionally, it is necessary for the customer to fabricate jigs or small items of test equipment to facilitate maintenance or repair. Full information should be provided along with applicable fabrication drawings.

LOWEST LEVEL OF SERVICE

On the initial failure of the equipment to function, the technician should have a set of guidelines to follow in order to restore the equipment to operational status in the minimum of time. This is best served by a fault diagnosis chart (Figure 7-5) that will, in most cases, lead directly to the main subassembly at fault. From that point on the service technician may:

a.   Replace the faulty item with a serviceable spare, realign or adjust the spare if necessary then return the equipment to operation, or

b.   If no spare assembly or module is available, effect the repair directly. This action, however, increases the equipment's down-time and requires that the service technician work quickly and effectively to pinpoint the faulty component, locate a spare to replace it, then return the equipment to operation.

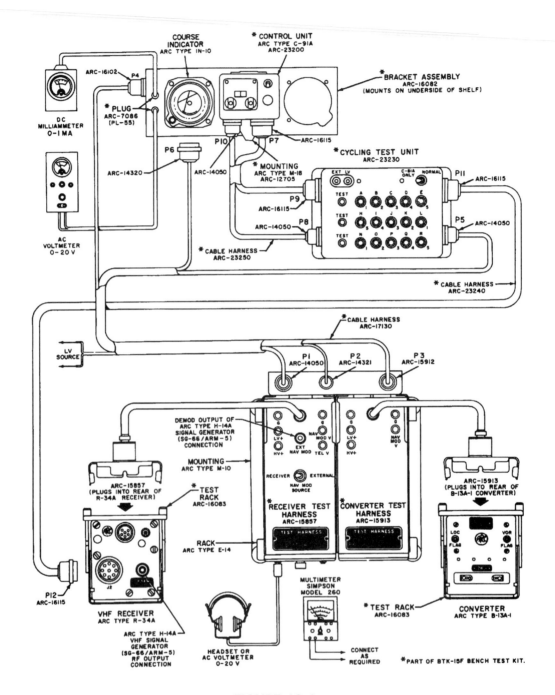

FIGURE 13-1
Servicing Test Setup

The inclusion of built-in test equipment (BITE) and other self-testing functions is making the diagnosis of problems significantly more easy today. On the other hand, the complexity of equipment is escalating as fast as the advances in modern technology.

Even in this age of the self-testing machine and disposable module/ component philosophy, good technical descriptions and accurate schematic diagrams are absolutely necessary for the service technician to fault-find and locate the "disposable" part in the equipment.

INSPECTION SCHEDULES AND PROCEDURES

Any equipment that has mechanical components should have regular inspections to ensure that parts are still in working order. Failure to do so jeopardizes the continued operation of the equipment in a safe and efficient manner. This is especially important with equipment that is located far from normal traffic areas and is operated remotely. Some areas of concern that may need to be included in a regular inspection schedule may include:

a. The condition of pulley belts on electric motors.

b. The tightness of connectors on coaxial cables.

c. Deterioration/corrosion of painted or plated surfaces.

d. Lubrication of moving parts.

e. Electrolyte levels in battery banks.

f. Cleanliness of filters in ventilation systems.

g. Condition of obstruction or safety lighting.

h. Looseness or unravels in guy wiring.

i. Excessive wear in gears or sliding components.

j. Deterioration of rubber components.

k. Overheating of power components or wiring.

l. Buildup of dust in equipment in certain environmental situations.

The frequency of inspections should be determined and a set of schedules planned to meet these needs. Generally, inspection schedules are carried out on:

a. Time schedules, that is, *yearly, monthly, weekly, daily*, or

b. Hours-of-operation schedules. This method requires some form of operating time counter and would be arranged in the manner of routine vehicle servicing, that is, 100's or 1000's of hours between inspections.

PERIODIC AND PREVENTATIVE MAINTENANCE

Although similar in nature to the periodic inspections, this activity performs specified tasks on the equipment at designated times, whereas the former procedure routinely checks conditions and takes action only if necessary. The two procedures should work hand-in-hand.

For example, precision equipment that includes mechanical motion or is prone to vibration should be subjected to both regular inspections and preventative maintenance. This can range from simple procedures such as checking and adjusting mechanical tolerances to the replacement of basic consumable items such as filters (these could be checked regularly and cleaned if necessary, but completely replaced after a certain period of time).

Bearings and rubber drive belts would be checked regularly for unexpected wear but would still replaced after a certain number of running hours.

TROUBLE ANALYSIS

Equipment trouble analysis requires a systematic method of localizing a problem, first to the major unit in which the malfunction exists, then to a particular area or a functionally related group of subassemblies and finally to a defective part or parts.

Equipment troubleshooting, based on the normal operating procedure, will often help to determine which subassembly is at fault.

An well-organized trouble shooting chart provides an effective tool for the determination of a fault. It reduces equipment downtime and the risk of unnecessary, and sometimes damaging, probing into parts of the equipment not involved in the problem.

The decision to use either the trouble shooting chart or fault-location table may be made in accordance with the particular equipment or system operation.

Alternatively, they can be used in conjunction with one another; for example, a tree can locate the faulty module, then the chart can supply a more detailed fault analysis.

REMOVAL AND REPLACEMENT OF PARTS

The more mechanized a piece of equipment is, the more important this section becomes. Very few assemblies can be taken apart without some form of instruction.

The designer of the equipment hardware often has little thought for producing an assembly that is simple to take apart. Functionality and operation are the designer's main concerns. Consequently, screws often have to be hidden behind escutcheons, under mounting pads, and in other most unlikely spots. Not knowing where these fixing parts are hidden can often lead to the equipment being damaged by service technicians trying to force open an assembly.

Quite often, special precautions are necessary; springs may be ejected and lost because of the failure to understand the correct disassembly procedure. Loosening or removing an item unnecessarily may necessitate extensive realignment procedures later on. Screws, when removed, may be of differing lengths, and incorrect reassembly could cause problems.

Some procedures need to have safety warnings against such hazards as residual voltages (see the section on Safety Instructions), springs under compression, or fragile but very dangerous material (cathode ray tubes in monitoring equipment). Some assemblies may need to be aligned in a certain way prior to removal otherwise reassembly becomes immensely difficult.

An example of a disassembly procedure outlining some special precautions is outlined below. Reference should be made to Figure 13-2 to follow the procedure:

Replacement of Printed Circuit Board Parts

Step 1: Release the four snapslide fasteners on the Dynaverter and carefully lift from the R34A chassis.

Step 2: Remove the two black oxidized screws from each side of the dust cover. Slide the cover back to clear the front panel, then lift it up and off.

Step 3: Remove the eight black oxidized screws holding the bottom plate to the chassis. Lift the plate clear.

Step 4: Disconnect the 21-pin connector and the adjacent coaxial cable connector from the I-F/A-F chassis A2.

Step 5: Remove the three black oxidized screws used to attach assembly A2 to the back and right side of assembly A1. **Important:** Note that the screw removed from the right side is shorter than the other two screws.

Step 6: Unscrew the two black oxidized screws from the top bearing plate and the two black oxidized screws from the main center plate to release the front panel and assembly A1 from assembly A2.

Step 7: Pull assembly A2 straight back to remove it from the rest of the receiver.

Step 8: For access to parts in assembly A1, remove the side plate by unscrewing the 11 attaching screws.

Most of the parts in the I-F/A-F assembly circuits are mounted on two printed-circuit boards, one each on assembly A1 and A2. Parts on the small A2 board are made accessible by removing two screws at each end of the capacitor mounting bracket (see Figure 5-9), unsoldering the ground wire from A2C8, and pushing the mounting bracket up...... ***and so on!***

To reassemble the unit, reverse the procedure of Steps 1 through 8. Before replacing the cover, however, apply a thin coating of *Dow-Corning 4 Compound* to the top of the frame assembly.

It will be noted in this procedure that special attention is given to screw colors and, in one case, the screw length.

In most applications such as this, the accompanying drawing is vital, so that directions in the text are more clear.

The disassembly procedures should be laid out in descending order from the highest order assembly down to the smallest practical part.

In the above sample description, the instructions for gaining access to the printed circuit board in the A1 assembly would appear next. Any special instructions for the replacement of certain parts on the printed circuit boards would follow.

Access to the A3 Tuner Assembly, seen on Figure 13-2, would be the next set of instructions and would refer to Steps 1 to 7 of the previous procedure.

ADJUSTMENT AND ALIGNMENT

Following most types of repair, some form of adjustment and/or realignment is necessary. Generally, the procedures to reset the equipment back to normal are the exact ones used in the initial setting up. There are, however, instances when the repair has not disturbed settings beyond a certain level, in which case the adjustment/alignment procedure is less complex.

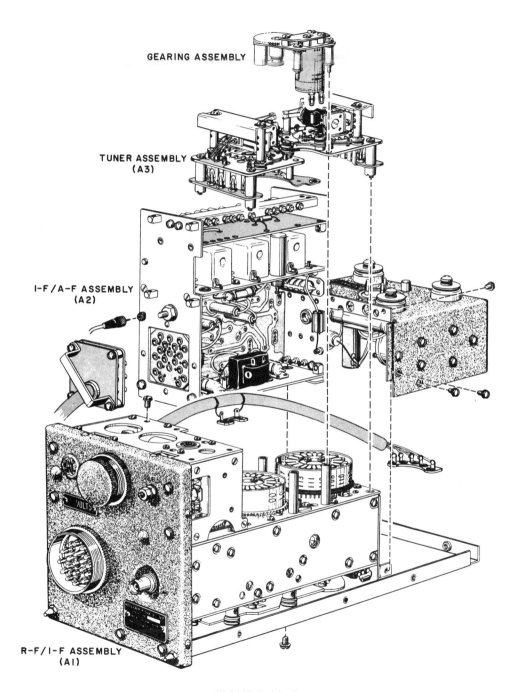

GEARING ASSEMBLY

TUNER ASSEMBLY
(A3)

I-F/A-F ASSEMBLY
(A2)

R-F/I-F ASSEMBLY
(A1)

FIGURE 13-2
Disassembly Illustration

The adjustment and alignment procedures should be set out in much the same way as for disassembly of the equipment, that is, a section detailing actions on a step-by-step form. However, it is usual to provide details of a test setup for performing some of the procedures, as typified in the following:

> **Adjustment and Alignment of R-34A Receiver**
>
> **General.** The following paragraphs describe the alignment and adjustment procedures for the R-34A Receiver. Table 5-1 lists the equipment required for the procedures. The test bench setup using ARC type BTK-15F Bench Test Kit is shown in Figure 5-24.
>
> Make the following test and adjustments before beginning the alignment of the R-34A receiver.
>
> *Step 1*: Interconnect the test equipment and Type 15F units as shown in Figure 5-24.
>
> *Step 2*: Apply power to the equipment and allow a 5-minute warmup period.......... ***and so on!***

The test bench setup figure mentioned above is shown in Figure 13-1.

PERFORMANCE CHECKS

When a measure of the equipment's performance is necessary, usually following maintenance or when the equipment has reached a set time limit, a set of performance checking instructions with an accompanying set of expected figures and tolerances, set out in table form, is required. The method of testing used can be detailed in the manner of those outlined in the last section, or it can be a set of brief tables with headings such as *Parameter*, *Switch Position*, *Normal Indications*, or *Level*.

Performance checking is carried out under near normal operating conditions and, in some instances, the results are recorded on a record sheet for historical comparison and assessment.

DRAWINGS

The drawings in the maintenance section of a manual form the bulk of all illustrations. They comprise:

a. Schematic, wiring, and flow diagrams of all parts of the equipment to accompany the theory of operation, printed circuit board layouts, and so on.

b. Setups and arrangements for maintenance, alignment, and performance checks.

c. Special illustrations such as the location or removal of parts, adjustment procedures, or lubrication points.

CHAPTER 14

ILLUSTRATED PARTS BREAKDOWN

INTRODUCTION

The compilation of a comprehensive illustrated parts list can be complex and time consuming; however, with the advent of computerized parts control, the cataloging of parts is now considerably easier.

Each unit of the equipment and any accessories should be covered in separate sections. Each section should contain an illustrated assembly parts list, a *part number numerical* index, a *reference designation* index, and, where applicable, a *cross-reference* to U.S. MIL- part numbers.

Each sectional parts list should be broken down into subassemblies, and detail parts. Items made from bulk stock, such as cut lengths of wire, tubing, and insulation material are not usually included.

The parts list *description* column lists each assembly or subassembly and its detail parts properly indented to show their relationship to the assembly or subassembly. Attaching hardware items (e.g., screws, washers, nuts) are also listed in the same indented column, immediately following the parts they attach. Some form of symbol should be used to show the completion of that assembly (a black line or a row of asterisks).

Parts that are limited in use to a specific model or serial number group are identified by noting the applicable model or serial numbers.

Code and part numbers of manufacturers, other than the parent company, should also be included in parentheses in the *description* column.

Except for attaching hardware, the quantities listed in the *units-per-assembly* column are the quantities per assembly or subassembly. The quantities therefore are not necessarily the total for the unit. Total quantities are listed in the unit's numerical index. An example of a descriptive parts list and the corresponding exploded view of the assembly are shown in Figures 14-1 and 14-2, respectively.

NUMERICAL INDEXES

The numerical index is located immediately following the main parts list for a particular assembly. It consists of a listing of the parts included in the unit parts list, arranged in alpha-numerical order. Then follows the figure number, reference column, and the total *quantity used* column (see Figure 14-3). This list provides both a handy part-to-location guide and a means of assessing the spares requirements of each assembly.

Parts that do not have part numbers are listed by item name and identifying characteristics. The *total quantity* column in the unit numerical index lists the total quantity of assemblies or parts for the unit. When a listed component is part of a higher assembly that is already listed, the annotation "REF" (Reference Only) is used. The letters "AR" (As Required) should be used to indicate indeterminate quantities (e.g., lengths of cable or wire).

REFERENCE DESIGNATION INDEXES

The unit reference designation index is a further index that lists the reference designations of components that are generally replaceable, cross-referenced to the applicable figure and index numbers and the part number (Figure 14-4). Where no figure or index number exists for the part, the figure or index number of the next higher assembly is listed, enclosed in parentheses; the part number of the next higher assembly is then listed. All reference designations used in illustrations and publications pertaining to the unit should be listed. The sequence of the listing should be alpha-numerical as before.

MANUFACTURERS' CODES

A listing of manufacturers should be compiled and a specific code assigned to each. These codes should be included in each part listing so that each item can be referenced to the manufacturer (Figure 14-5).

Figure & Index No.	Part Number	Item Description	Units Per Assembly
2–	22500	CONTROL ASSEMBLY	REF
–1	25886	• LIGHT ASSEMBLY, Panel	2
–2	—	•• LAMP HOUSING	1
–3	—	••• GROMMET, Panel Light	1
–4	—	••• SPRING, Lamp Retainer	1
–5	—	••• HOUSING, Lamp and Filter Assembly	1
–6	8622	• LAMP, T-1¾, Flange Base, 28V, 0.04A	2
–7	16331	• KNOB, Control	2
–8	306-0012	• SETSCREW, Cup Point, 6-32 UNC-2A, 0.18 Lg, Sst(AP)	4
–9	20403	• KNOB, Control	2
–10	306-0016	• SETSCREW, Cup Point, 6-32 UNC-2A, 0.25 Lg, Sst (AP)	4
–11	23588	• PANEL, Plastic, Engraved	1
–12	113-0024	• SCREW, Machine, Bnd Hd, Blk Brs, 3-38 UNC-2A, 0.38 Lg (AP) .	2
–13	10353	• WASHER, Spring, Helical, No.3 (AP)	2
–14	25886	• PLATE, Identification	1
–15	8956-2010	• SCREW, Tapping, Thread Forming, Rh, Blk, No.2, 0.16 Lg (AP) . .	2
–16	23721	• COVER	1
–17	23591	• PLATE, Mounting	1
–18	503-0012	• SCREW, Assembled Washer, Brs Screw, Blk, Bnd Hd, 3-48 UNC-2A, 0.18 Lg, Sst Lock Washer	4
–19	12097	• CONNECTOR, Receptacle, Electrical	1
–20	5130	• NUT, Round, Knurled, 1⅛-24 Thd by 0.109 in. (p/o -19) (AP) . . .	1
–21	12355	• CONNECTOR, Receptacle, Electrical	1
–22	5131	• NUT, Round, Knurled, 1⅝-24 Thd by 0.109 in. (p/o -21) (AP) . . .	1
–23	12357	• CONNECTOR, Receptacle, Electrical	1
–24	5131	• NUT, Round, Knurled, 1⅝-24 Thd by 0.109 in. (p/o -23) (AP) . . .	1
–25	—	• RETAINER, Lamp Housing (p/o -1)	2
–26	—	• NUT, Hex (Part of -1) (AP)	2
–27	—	• WASHER, Lock (Part of -1) (AP)	2
–28	23730	• RESISTOR, Variable, Comp, 5kΩ, ±10%, 2W	1
–29	4697	• NUT, Round, Serrated, Brs, Ni Pl, ⅜-32 Thd (AP)	1
–31	23115	• RESISTOR, Variable, Comp, 100kΩ, ±10%, 2W	1
–32	4697	• NUT, Round, Serrated, Brs, Ni Pl, ⅜-32 Thd (AP)	1
–34	12049	• TERMINAL	1
–35	153-0016	• SCREW, Machine, Fl Hd, Brs, Blk, 3-48 UNC-2A, 0.25 Lg (AP) . .	1
–36	4103	• WASHER, Lock, Helical, Sst, No. 3 (AP)	1

FIGURE 14-1
Parts List Page

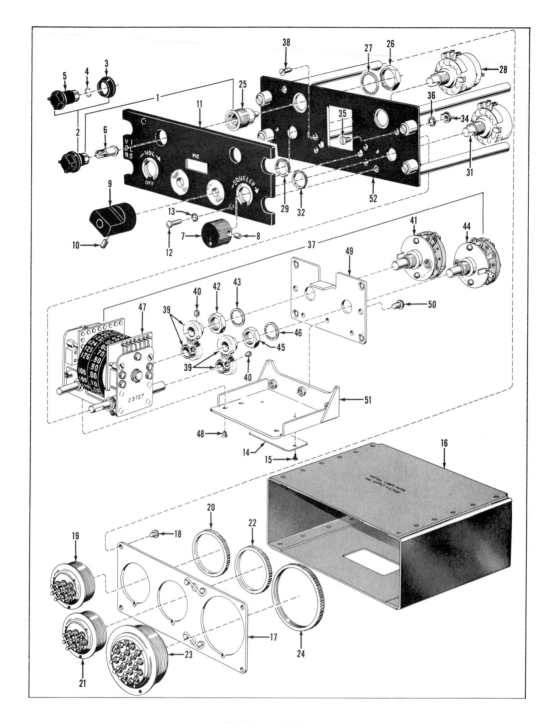

FIGURE 14-2
An Illustrated Parts Breakdown

MILITARY IPB PREPARATION

When a company is involved in the preparation of technical manuals destined for use by the US government, especially a branch of the armed forces, the layout requirements of the IPB become more stringent. As with other major sections of the manual set (Description, Operation, Maintenance, Overhaul, etc.), the IPB is usually prepared as a separate book and required to conform to a specific military specification. A typical example is MIL-M-38807, which details IPB preparation for the US Air Force.

This military specification spells out in great detail the exact requirements of every aspect of IPB manual preparation. Much of the basic IPB layout described in this chapter follows the general requirements of a military specification.

NUMERICAL INDEX

Part Number	Figure & Index No.	Total Quantity	Part Number	Figure & Index No.	Total Quantity
Gearing Subassembly	2-44	1	200-0101	5-18	1
Interconnecting Box	6-129	1	200-0103	6-50	1
Lever Subassembly	4-23	2	200-0104	6-86A	1
	4-33		200-0105	3-11	1
Locknut	2-20	8	200-0106	3-57	1
	3-2		200-0107	5-17	1
	5-21		200-0108	6-14	6
Plate Subassembly	4-53	1		6-52	
Panel Subassembly	3-32	7		6-84	
	3-117			6-46	
	3-123		200-0109	6-11	1
	5-5		200-0111	5-58	1
	5-20		200-0112	6-58	1
	5-42		200-0113	6-15	1
	5-62		200-0114	5-53	3
BS48179	2-4	REF		6-17	
BU20251MV5V	3-90	REF	200-0201	3-55	1
BU25272NX5Z	3-21	REF	200-0202	5-40	1
CB1011	5-18	REF	200-0203	6-13	3
CB1031	6-50	REF		6-60	
	6-66A			3-54	
CB1041	3-11	REF	200-0204	5-39	1
	3-57		200-0205	6-32	2
	5-17			6-48	
	6-14		200-0206	6-11A	4
	6-52			6-15	

FIGURE 14-3
A Numerical Index of Parts

REFERENCE DESIGNATION INDEX

Reference Designation	Figure & Index No.	Part Number	Reference Designation	Figure & Index No.	Part Number
C201	4-66	8617-0201	J201	4-57	13152
C202	4-27	8966-0210	J202	4-58	13152
C203	4-28	27166-0106			
C204	4-51	28500-0104	K201	4-39	16665
C205	4-52	27155-0203	K202	(4-102)	p/o 16667
C206	3-8	27155-0203	K203	4-14	16665
C207	4-2	16143			
C209	4-60	27155-0203	R201	4-69	201-0105
C210	3-12	851774	R202	4-63	201-0514
C211	3-12	8716-0201	R203	4-67	201-0111
C212	3-12	21485-9101	R204	4-23	201-0105
C213	3-12	27155-0203	R205	4-85	201-0105
C214	3-12	16143	R206	4-79	85670
			R207	4-23	201-0106
CR201	4-18	166600	R208	4-25	85733
CR202	4-18	166600	R209	4-86	201-0105
CR203	4-18	166600	R210	4-87	201-0111

FIGURE 14-4
Reference Designation Index

MANUFACTURERS' CODE INDEX

Name and Address	Code
Allen-Bradley Co., Milwaukee, WI	AB
American Brass Co., Waterbury, CT	AR
Cinch Manufacturing Corp., Chicago, ILL	CIN
Cornell-Dubilier Electric Corp., South Plainfield, NJ	CLD
Clifton Precision Products Co., Clifton Heights, PA	CLIP
Communications Engineering, Chicago, ILL	CMMU
Colin Campbell Co., Inc., Danbury, CT	CMPB
Centralab Division of Globe-Union Inc., Milwaukee, WI	CN
Dale Products Inc., Columbus, NE	DABU
Electro Motive Mfg. Co., Willimantic, CT	EMM

FIGURE 14-5
Manufacturers' Code List

CHAPTER 15

APPENDIXES AND ADDENDA

INTRODUCTION

Appendixes and addenda are used to present information that cannot, or should not, be incorporated into the main body of the manual. By omitting certain kinds of information from the main text sections and placing them in an appendix, the readability of the manual may improve considerably. However, it is not uncommon for authors to locate material in an appendix simply for convenience rather than include the material into the main body.

The difference between an appendix or attachment, as it is sometimes referred to, and an addendum is not clearly defined. This author makes this distinction, based on personal preference only.

THE APPENDIX

The *appendix* is located at the rear of the manual and may contain the following:

a. Procedures and actions that may be performed occasionally or over long time intervals. This keeps large or complex descriptions or procedures away from the general, more frequent, activities. For example, an instrument may have a special tuning procedure that is performed only when a special component is replaced due to age or failure. With this procedure located away from the day-to-day operational procedures, it does not force the service technician to keep turning past this block of instructions to move on to more routine operations.

b. Sometimes a special set of installation instructions is provided for equipment. Locating this set within an appendix provides the complete information, always available for reference, but without cluttering up the main text.

c. Data sheets containing readings based on the original installation or initial turn-on of the equipment. These would provide a reference for comparison of the equipment's performance as it ages.

d. Specimen material, that is, examples of performance figures under certain conditions, sample calculations, and so on, not necessary to the understanding of the text but desirable from the point of view of understanding the philosophy of the system.

THE ADDENDUM

The *addendum* is a self-contained document in the form of a mini-manual that provides all the necessary information about an optional item (see Figure 4-2) that may or may not have been fitted during the equipment's construction or retro-fitted after initial installation. This document is usually located at the rear of the section pertaining to the specific assembly to which the optional subassembly will be fitted.

As seen in the structure chart in Figure 4-2, the addendum contains theory, parts information and all relevant drawings. The references used in this drawing, (i.e. **C5A**) represent a **C**ommon module for the Part **5** subassembly (Local Control Unit) with **A** being the first of several possible addenda.

For example, a vehicular radio-telephone (mobile two-way radio) could be converted to act as a base station. It would need a special desktop housing, mains-operated power supply, and an antenna system. This optional equipment, when purchased, would come complete with the addendum to be added to the existing mobile equipment handbook. Thus all aspects of the optional equipment would be included, permitting operation, maintenance, and repair functions to be performed.

The major difference between these two document types is that the appendixes are referred to within the main body of the manual. The addendum is not, because its inclusion is dependent on the customer's requirements before, during, or after the equipment is ordered. However, a reference to options likely to be fitted may be made either in the front matter of the manual or in the section pertaining to the assembly where it would be fitted. A tabbed divider would suffice to indicate to the manual user that special additional information has been added to the specific section.

INTEGRATION

The use of appendixes must be such that the reader does not have to disrupt a procedure being performed to locate something in an appendix, perform an additional specified task, then return to the original procedure.

In such instances it is preferable that a procedure which encompasses an appendix procedure, be completely contained within that appendix. For example, a procedure that deals with the adjustment of a device may refer to an additional set of adjustments in an appendix if certain conditions exist. Rather than forcing the user to jump back and forth between the two sets of text, the complete adjustment procedure should be located within the appendix. Now, if certain conditions exist, the appendix procedure will be applicable; if not, the procedure in the main body of text will be used.

LAYOUT

The layout of an appendix should follow that of a major manual section; that is, it should start on the right-hand page and have a cover sheet or an appropriately headed first page. The reference number and title of the appendix should be displayed prominently. The cover sheet does not usually include a page number, but a first page that uses header references only should be numbered. A good rule of thumb for appendix preparation is to set it up so that it can be removed from a manual and still be readily identified as a appendix to that specific manual.

If the appendix information is lengthy, it may be necessary to include an index, either as a separate page or as the first section of the front page.

REFERENCING

Appendixes should be referenced with letters only, for example, Appendix A or Appendix B. Paragraph numbering and table/figure numbering can be prefixed with the specific letter, for example, Section B3.6.1, Figure B1-5, and Table B5-1 (all part of Appendix B). This system is helpful, if regular reference to appendix material is made in the main manual, to reduce confusion in understanding which material is referred to.

Separate appendixes can be used for each subject, and each should have a separate title and identification number, if appropriate.

Figure 15-1 shows a sample appendix format (first page), with table of contents and text on the same page. Paragraph numbering is conventional.

APPENDIX A

ELECTRICAL HEAT TRACING SYSTEM

Table of Contents

1.0 INTRODUCTION

2.0 PURPOSE

3.0 DEFINITIONS

4.0 OPERATING PRINCIPLES

4.1 General
4.2 Operation
4.3 Mechanical
4.4 Redundancy
4.5 Operation in Hazardous Areas

5.0 THERMAL INSULATION

5.1 Insulating Materials
5.2 Weatherproofing
5.3 Vertical Runs
5.4 Below Grade Insulation
5.5 Water Drainage Pipes

6.0 HEATING NEEDS

6.1 Heat Loss of Pipes
6.2 Other Factors

7.0 SELECTION OF CABLES

7.1 Introduction
7.2 Series Heat Tracing Cable
7.3 Parallel Wattage Heater Cable
7.4 Parallel Self-Regulating Cable
7.5 Cable Selection

8.0 POWER SUPPLY

8.1 Power Sources
8.2 Ground Fault Protection
8.3 Wiring of Series Heater Cable
8.4 Wiring of Parallel Heater Cables
8.5 Ammeters

9.0 TEMPERATURE CONTROL

9.1 Types of Controllers
9.2 Location of Sensors
9.3 Temperature Set Points

10.0 ALARMS

10.1 Types of Alarms
10.2 Temperature Annunciations

1.0 INTRODUCTION

This appendix provides general information and recommended practices for the installation of electrical heat tracing on mechanical and instrumentation piping (or tubing) systems in arctic environments.

2.0 PURPOSE

The purpose of this appendix is to provide information on electrical heat tracing supplied with the installation of the equipment so that wiring or instrument air systems are available for use during normal station operation, or in an emergency during shutdown.

The function of an electric heat tracing system is to provide, to an insulated mechanical piping system, heat that will maintain the system within its specified operating temperature range. The mechanical and electrical properties of the heating system should not be impaired by continuous or intermittent operation under the environmental and operating conditions specified.

3.0 DEFINITIONS

The following definitions related to electrical heat tracing apply. More definitions can be found in the IEEE standards listed as references.

Freeze Protection: The use of electric heat tracing systems to prevent the temperature of fluids from dropping to or below the freezing point of the fluid. Freeze protection is usually associated with piping, pumps, valves, tanks, instrumentation, etc, that are located outdoors or in unheated buildings.

Heat Sink: A part that absorbs heat. Heat sinks, as related to electric heat tracing systems, are those masses of materials that are directly connected to mechanical piping, like valves, tanks, pipe hangers, etc, that can absorb the heat generated by heaters, thus reducing the effectiveness of the heater.

Parallel Heating Cable: A cable including heating elements that are electrically connected in parallel, either continuously or in zones, so that watt density per linear length is maintained irrespective of any change in length for the continuous type or for any number or discrete zones.

Process Control: The use of electric heat tracing systems to increase or maintain, or both, the temperature of fluids (or processes) in mechanical piping systems in power generating stations.

Series Heating Cable: A cable incorporating one or more continuous fixed resistance elements connected to provide a fixed rated output at a given voltage, temperature, and length.

A-1

FIGURE 15-1
Appendix Front Page

CHAPTER 16

AMENDING MANUALS

THE NEED FOR AMENDMENTS

The decision to subject a technical manual to amendment or revision after issue is based on several factors:

a. Scope and use of the equipment and the level of detail involved; for example,

    (1) A domestic snowblower operation and maintenance handbook—minimal technical information; follow-up information most unlikely or necessary,

    (2) A waveshape analyzer manual — a high cost, complex instrument with detailed procedures. The manual would have to be accurate and up to date, so amendment procedures would be necessary.

b. If the equipment is of a type where errors in the handbook are likely to cause difficulties in user comprehension, leading to incorrect operation and subsequent damage to the equipment.

c. Possible malfunctions of the equipment due to incorrect maintenance, and the extent to which such malfunctions would affect operation.

d. Possibility of personal injury due to procedures not clearly defined, or warnings not provided against the existence of hazardous situations.

INCORPORATING AMENDMENTS

The amendment of a technical manual by the substitution of pages or the incorporation of handwritten changes appears to be an easy and foolproof operation, to the originator. However, unless the originator is very careful and explicit in the instructions given, the manual holder may become frustrated and fail to incorporate the vital changes, with disastrous results.

The amendment package should indicate the following to the manual user:

a. A brief reason for the changes.

b. Exactly what pages are to be removed and/or replaced.

c. The disposal of the old pages.

d. A reminder to complete the amendment block.

A typical amendment instruction would be written in the following manner:

1. Page iii, amendment certificate. Replace new page iii with updated amendment block page. Fill in date and signature when amendments have been incorporated.

2. Chapter 6: Remove and destroy pages 6-1 through 6-6 inclusive. Insert new pages 6-1 through 6-6.

3. Chapter 16, page 16-8, line item 22. Change column 3 (QTY) from 5 to 4.

4. Chapter 18, Figure 18-4. Delete notes 6 and 7.

The first amendment instruction inserts a new revision block incorporating these amendments. The user has to complete the block on finishing the amendment set. The next amendment changes pages in which either a number of text changes occurred or it was considered too difficult to make handwritten changes. Amendments 3 and 4 are small enough to warrant hand-inserted changes, so replacement pages are not necessary.

DISTRIBUTING AND RECORDING AMENDMENTS

Preparing amendments to manuals is of no value unless the locations of all manuals are known. Generally, manuals of the type that are to be revised are in the hands of customers and/or some responsible person or department that can be contacted for equipment and document changes.

In large organizations such as civil aviation or maritime bodies, where there may be a dozen widely scattered equipment installations, manual revisions/amendments are often forwarded to a central documentation department, where some level of internal control of manuals is exercised. In these organizations, some form of feedback is often required to prove that the amendment has been incorporated, such as a certificate that has to be completed by the manual holder and returned to the distributing department on incorporation of the revisions.

AMENDMENT IDENTIFICATION

There are several ways in which an amended page or section of text can be indicated to the reader:

a. A vertical black bar on the outside margin, extending the length of each paragraph altered, as shown immediately to the right, and/or

b. A notation in small text on the very bottom of the page (as shown) or in the margin between the binding holes, as displayed to the left, indicating the revision number and effective date, or

c. A revision block (usually on a large drawing) with changes detailed.

The status of a manual with respect to its amendment level is indicated by a revision block located close to the front of the book, preferably on the same page as a copyright statement. This block is either filled in by hand or is routinely upgraded with each amendment issue. The latter is the most efficient because it ensures that the correct information is always inserted. A typical revision block is shown in Figure 16-1.

REVISION NUMBER	EFFECTIVE DATE	DATE OF AMENDMENT	COMPLETED BY (SIGNATURE)	AMENDMENT DETAILS
1	13JUN93	30JUL93	*L. Oatemitten-Lee*	Replaced pages 6-1 to 6-6 with new set. Deleted notes 6 and 7 on Figure 11-3. Deleted line item 22 on page 16-8. Amended line 22 QTY column to 4.
2	27DEC93	13MAR94	*S. M. Jackson*	Replaced Figure 18-2 with new.

FIGURE 16-1
Revision Block with Typical Entries

AMENDMENT CONTROL

When a manual is subject to amendment, there is always the risk that some copies will remain unamended or only partially amended. Out-of-date copies are a source of confusion and could result in damage to equipment or personnel if an important amendment was not incorporated. To reduce such risks, amendments should never be issued in a haphazard manner. Except in an emergency, when immediate action is necessary, amendments should be issued at specific times, for example, every six months or once a year, although it is necessary to follow up the incorporation of any important equipment modifications with new text/drawings as they occur, irrespective of the timing.

From the time the manual is first printed, one copy should be maintained by the authors or their department to record all changes pertaining to the manual. These changes will be derived from company personnel, using the manual for training, repair, or installation tasks; feedback from customers; or from the engineering staff as they continue to upgrade the equipment.

Often, where several different pieces of equipment or several models of the same piece of equipment from a single manufacturer exist in an organization, there could be some confusion as to which manual a specific set of amendments apply. Where such a possibility exists, the manual preparation group should consider providing each manual set with a reference number. This number should be clearly visible on the front cover and spine of all binders in the set.

Using the instrument landing system equipment as an example once more, we could come up with manual reference numbers ILSGP74, ILSLC74 then ILSGP82 and ILSLC82. The last two digits indicate either a 1974 model equipment or a later 1982 model. The GP and LC notations identify the two major units in each installation, that is Glidepath and Localizer, as each would have its own manual set. These reference numbers are virtually self-explanatory and readers have no difficulty in determining which is which.

So an amendment set applicable to handbook ILSGP82 would be accurately directed to the correct handbook, namely, the ILS Glidepath handbook for the latest 1982 model.

CHAPTER 17

PREPARING CAMERA-READY COPY

INTRODUCTION

Prior to the revolution of the computer and desk top publishing, the manual producer had to deal with the technique of cut-and-paste to prepare a camera-ready manual. This technique involves the location and alignment of printed sections of text and various graphics onto specially prepared layout cards. These cards were in the form of an individual page of a manual or book and contained various guide lines in light blue ink, which made them photo invisible.

Today, the personal computer has replaced almost all of these laborious time-consuming tasks; however, some paste-up is still occasionally required, mainly in the illustration field.

TEXT PAGE PREPARATION

Laying out the manual pages on a personal computer is a matter of using specially-prepared templates (page forms) as found in such desktop publishing packages as *PageMaker, Quark XPress*, or *Ventura Publisher*. While such software programs require trained personnel to run them, the techniques needed to prepare the manual are fairly straightforward.

If, on the other hand, the company does not use any of these software programs, preparing suitable templates on word-processing programs such as *Microsoft Word* or *WordPerfect®* is relatively easy for someone skilled in any of these programs. *WordPerfect®* was used to prepare this handbook. In this instance, three special templates were prepared; a chapter heading page, a right-hand page, and a left-hand page.

Each had crop marks (small alignment guides for setting up and cutting) on the corners (for the smaller than 8.5 × 11-inch page sizes), title heading, page number, and a horizontal line. All tabs, margins, and typefaces were preset.

Text and illustrations were prepared independently and loaded onto the appropriate page as the manual progressed. The preparation of a detailed manual structure (Chapter 4) was important to the procedure. An example of a standard two-column manual page with a table and graphic inserted is shown in Figure 6-1. When constructed in this form and laser-printed on good quality paper, it constitutes a suitable camera-ready page for either photocopying or offset printing reproduction.

PREPARING ILLUSTRATIONS

In any illustrated technical manual, the highest cost by far in most cases is the cost of preparing the illustrations. There is an adage used in aircraft industry manual publishing that makes good sense: *If the text can explain it—don't draw it!* Review the text carefully and then ask: "Is this illustration really needed?"

Some factors to consider when preparing illustrations for a manual are the following:

a. Avoid redrawing wherever possible. Search for existing drawings that may have all or part of the required drawing already.

b. Make illustrations as simple as possible, eliminating all unessential details. If using existing engineering drawings, be prepared to cut out unwanted sections, then edit, crop, and so on, to get all or most of the required illustration. Remove such material as borders, revision blocks, and parts lists.

Illustrations for manuals can be prepared in several ways, each depending on the size of the drawing being handled:

Small illustrations: Scanned and converted to suitable graphic files for direct incorporation into the pages of the manual. These may be positioned on separate pages or combined with the text as seen in Figure 6-1.

Illustrations that are no more than about 60 percent larger than the finished size can be reduced on a good-quality photocopier (using high quality paper) to a size that will fit the scanner, then converted to graphic files.

If the text size on a reduced illustration is marginal, and there is not too much of it, a satisfactory solution is to retype the wording in a larger, clearer font and paste the new text over the old. This, however, has the disadvantage that both a copier and a scanning device will produce a shadow mark along the leading scanned edge of the pasted-in text. This can be fixed easily by a liberal coating of typist's white correction fluid along that edge. This removes the step and the scanned copy should be clean.

Figure 17-1 shows a *before* and *after* treatment of a mechanical drawing using this technique. The original is too cramped and contains material not needed for a manual. It has been removed simply by cutting away as much extraneous material as possible, then using correction fluid to blank out arrow lines within the body of the piece. In some places, sectioning lines have been redrawn.

On a relatively large original, the cutting and blanking rework can produce a very fine illustration at the expense of an hour or so of careful work. In some instances, because of the original drawing size, it may not be practical to use a scanner or photocopier. The finished illustration from a large original may have to be an 11-in. by 17-in. foldout. In such a case, the solution is to have a photomechanical transfer (PMT) produced for a nominal fee by an external reproduction company. The PMT is a crisp, clean, and shadowless print on light nonglossy photographic paper. The PMT can be pasted onto an 11 in. by 17-in. layout sheet with the necessary captions and page number added.

A layout sheet can be made larger than the finished 11 in. by 17-in.: all text and illustrations are pasted in, then the layout sheet is reduced onto a PMT at the exact finished size. This is a useful exercise when the amount of material adhering to the layout sheet is large and too delicate to be subjected to normal handling. It should be remembered that any larger than normal layout sheet requires any text, captions, and so on, to be in the correct proportion, so that on reduction, it is at the size of the rest of the pages.

When dealing with reduced material, cut-and-paste materials, and layout work, an invaluable tool is a *proportional scale*. It is essentially a circular scale that provides percentage changes between any two values. For example, you wish to reduce an original drawing, on a copier, from 5½-in. by 4½-in. to fit into a 20-pica wide column (which is slightly wider than 3¼ in.). Using the proportional scale, you find that the reduction percentage is 67 percent and the height of the drawing will reduce to 3 in. With most modern photo-copiers now providing fine incremental enlargements and reductions, this tool is invaluable. A readily available proportional scale is made by the C-Thru® Ruler Company and can be purchased through any major art supply house in the United States and Canada.

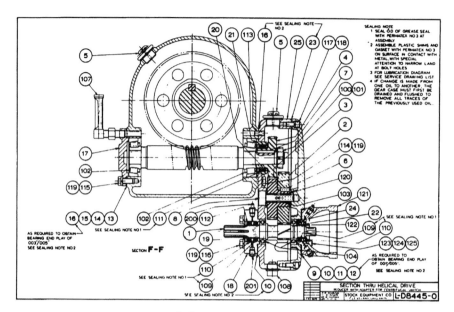

FIGURE 17-1(a)
Original Illustration

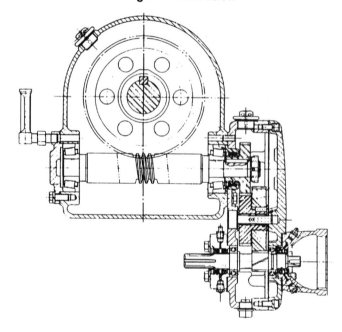

FIGURE 17-1(b)
Illustration After Rework

CHAPTER 18

PRINTING AND BINDING

INTRODUCTION

The printing and binding of the large sheaf of text and illustrations that comprise the vital organs of the technical manual is almost as important a phase of the production process as any other. Although a technical manual is not presented in the same manner as a book that is intended to go on sale at booksellers, it gets more day-to-day use than probably any other type of book, even more than the reference work and academic textbook.

This chapter covers the essentials of paper types and the various binding procedures available today. While most of it will not have any direct application for the manual producer, an understanding of the more common paper and the way in which it can be bound should prove to be of some value.

PRINTING PAPERS

Selection Criteria

The nature of the technical manual itself has a great deal to do with the choice of paper. Answers to the following questions often lead to the type of paper to be used:

a. Are there to be foldout pages?

b. Under what climatic conditions will it be used?

c. Will it be subjected to extreme handling or soilage?

d. Does the manual contain any color illustrations?

e. Is the company image or an impression of high quality to be conveyed?

f. Will the production run be a long or a short one?

g. Are there likely to be reprints of the manual?

h. What printing process is the most suitable for this job?

Paper Grades

Fine paper may be classified as bond, book and text, cover, bristol, blank, and board. An understanding of paper types and grades is of value to those responsible for organizing the manual printing. Papers are normally described by finish, grain direction, the printing process for which they are designed, weight, and sheet size.

Bonds: Bond papers include parchment and ledger papers. They are usually watermarked and have a matt surface. They are used almost exclusively for office stationary, letterheads, envelopes, and business forms.

There are many grades, ranging from a content of 100 percent sulfite fibers to a mixture of sulfite and rag fibers and, in high quality brands, parchment and ledgers, to a content of rag fiber alone.

They also range in weight from onion skins and airmail papers to safety sheets, which are used for cheques, and crisp sheets used in ledger books.

Book and Text: Book papers provide readability and good ink receptivity for manual, book, periodical, and publicity printing. Text papers are special-use papers with attractive textures for use in high quality folders, brochures and booklets, and commemorative and promotional purposes.

Cover Papers: Used for book jackets, these papers are usually of heavier weight and their colors and finishes may match or contrast with the product they enclose or protect.

Generally they have a uniform printing surface, good folding characteristics, durability, and strength. The wide range available indicates their importance in presentations and printed promotional material. Cover stock can be regularly found in menus and announcement sheets. These are ideal for the production of slip-in title sheets and spine inserts for manual covers.

Offset Paper: This is similar to coated and uncoated book paper used for letterpress printing, except that it has been treated to resist the slight moisture present in offset printing.

Bristols, Blanks, and Boards: These are used for mailers, tickets, announcements, and file cards. The rigid multi-ply blanks are used for signs and stiffeners. Boxboard has both rigidity and folding qualities depending on grain direction and is printable on one side only.

Finish

The appearance of papers and the way they feel to the touch introduce certain other terms. Book papers that are not coated are classified from less smooth to smoother by the following terms:

Antique: Used for text type and average line work in illustrations.

Vellum: Used for text type and fairly open line work and screened copy.

Machine or Plate Finished: Various grades provide a fine quality paper for fine line or halftone reproductions.

Fancy Finished and Textured: Uncoated papers with a linen, woven, or other fancy finish.

Matte: Papers that are slightly rough to the touch.

Paper manufacturers have their own systems of designating paper grades and colors. Samples are usually available from the publishing or printing companies, or from the manufacturers themselves.

Grain

Grain is an important factor in both printing and binding. It refers to the formation of fibers in paper and is important in running a paper type through a press and in any subsequent folding that may be necessary.

Grain affects paper in the following ways and needs to be considered in the proper usage of paper: papers fold best with the grain direction. This avoids roughing and cracking of the fibers when folded across the grain and permits accurate folds. However, lightweight paper can sometimes be folded against the grain.

As the weight of a paper increases, it may become necessary to crease the paper or board before bending.

Weight

This can be a confusing parameter. Weight refers to the thickness of paper and is known in North America and Canada as *basis weight*.

Generally, paper is identified by ream (500 sheets) weight, for example, 20 pound bond or 70-pound uncoated.

For bond papers, basis weight is usually referred to as *substance*, a term often found on the label of photocopier and typing paper. Here the basic uncut sheet size is 17 × 22 inches, which will cut evenly into 8½ × 11-inch sheets. When cut and packaged, it may be labeled:

White Bond 8½ × 11 in. (216 mm × 356 mm) 500 Sheet Package — Substance 20 lb (75 g/m²) 10M. Grain long.

But paper is usually listed in size-and-weight tables and price lists on a 1000-sheet basis; 17 × 22–10M for 20-pound bond paper; that is, 1000 sheets of bulk size paper of size 17 × 22 inches, weighs 10 pounds (this is where the 10M on the above label comes from!).

Similarly, a package of 11 × 17 inch (279 mm × 432 mm) bond paper would be labeled identically to the example, except for the cut size.

But a ream of a different type of paper, because the basic uncut size is different, may weigh more or less. This uncut size varies from one category of paper to the next.

In the metric (SI) system, the bulk size of the paper is always 1 m² and its weight is expressed in grams per square meter (g/m²). Both size/weight parameters appear on Canadian-made paper.

Folding

The two common kinds of folds are parallel and right-angle. Parallel folding is, as the name implies, where each fold is parallel to the other. A prime example is a business letter that requires two parallel folds for insertion in an envelope.

Another type of parallel fold is the accordion or fan fold, used extensively for large diagrams in technical manuals. A right-angle fold is two or more folds, with each fold at right angles to the preceding one. Most commercial greeting cards are folded with two right angle folds.

Paper is usually folded on a mechanical folding machine. Each sheet is fed via an automatic feeder into the folder. The machine is adjusted to accept various size sheets, which can be printed with as few as 4 and as many as 64 pages (these printed sheets are called *signatures*). The sheet enters the folder and is mechanically folded in such a way that all pages are in their correct order and alignment but are joined at different sides.

Scoring, trimming, slitting, perforating, and pasting can be performed at the same time as the folding on these machines. These operations are generally inexpensive and time saving if care has been taken to set the signatures correctly.

In designing manuals to be bound, the different types of folds and the limitations of mechanical folding should be considered at the planning level. Otherwise, one or more folds might end up being a costly hand-folding operation. Figure 18-1 illustrates some common types of folds used in booklet preparation.

Types of Folds

Four-Page Folder: A single fold along either the long or short dimension and used for small advertising brochures, instruction sheets, price lists, an so on (Figure 18-1, fold types 1 to 4).

Six-Page Folder: Made with two parallel folds, either regular or accordion. Used for everyday business letters, circulars, and promotional folders (Figure 18-1, fold types 5 and 6).

Eight-Page Folder: Two parallel folds, shown as type 7, one parallel and one right-angle fold (type 8) or three parallel accordion folds (type 9). These folders can be bound into an eight-page booklet.

There are further fold sets producing 12-page and 16-page booklets, which are the maximum generally used. The number of folds is related to the size of the finished page. The smaller the page size, the more pages that can be printed on a single sheet of paper, and the smaller the number of printing operations needed to produce the pages for a book.

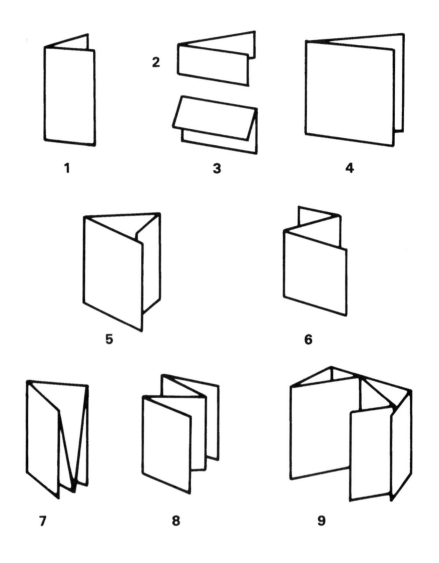

FIGURE 18-1
Common Types of Paper Folds

Collating

Once the signatures are folded, they must be gathered or collated in the correct sequence for binding. Collating can be done by hand or machine, depending on the size of the job.

The next operation is the actual binding of the signatures to form the book. There are various ways of achieving this, generally depending on the thickness of the finished book as well as the quality desired. The two basic methods are either *stitching* or *case binding* (also called perfect binding and sewn binding—see the following section on binding methods).

Stitching

There are two methods of stitching, saddle-stitching and side-stitching (see Figure 18-2). These are quick and economical methods of binding brochures, pamphlets and small manuals and involves the insertion of metal staples inserted through the crease or gutter made by the folding process. The thickness or bulk of paper determines the style to be used. Common magazine rack publications such as *People, Time* and *Newsweek* are usually saddle-stitched. The thicker *National Geographic* is side-stitched, then fitted with an outer cover.

SADDLE STITCH   SIDE-STITCH

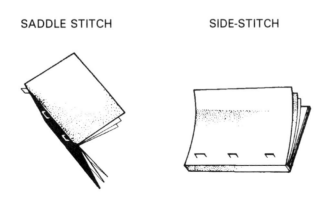

FIGURE 18-2
Stitching Styles

Saddle-Stitching: In this method, the magazine or booklet is placed on a saddle beneath a mechanical stitching head, and staples are forced through the backbone or spine.

This form of binding is the simplest and most inexpensive of the two types. Booklets will lie flat and stay open for ease in reading. Most booklets, such as theatre programs and new automobile advertising catalogs are saddle-stitched. Hand machines, essentially deep-reach heavy-duty staplers, are readily available for small jobs.

Side-Stitching: This is used when the thickness of the book is too great for saddle-stitching. The series of signatures are collated, and then stitches are inserted about ¼ inch from the back edge. Care must be taken to allow the inside margin to be wider than in a saddle-stitched booklet. This form of stitching also permits the binding of single pages and foldouts together with the folded sheets.

After stitching, the three open sides of the booklet are trimmed on a guillotine paper cutter to achieve even edges.

BINDING METHODS

A number of methods exist to bind together printed material. Some of the simplest examples are the familiar loose-leaf ring binding or the plastic bound and wire-stitched brochures. Frequently one or another of the folding methods shown in Figure 18-1 is used before binding. As in folding, binding may be done by hand or by machine, or by combining both hand and machine operations, depending on the complexity of the job.

Some examples of binding methods are given here.

Loose Leaf (Mechanical) Binding: Loose-leaf binding is used for most technical manuals because of the ease of assembling them, amending them, and having the added convenience of being able to remove material and relocate it into other three-ring binders. Special rings, called D-rings, can be used in binders so that the pages will have an almost square edge when closed or opened (Figure 18-3).

Loose-leaf binders can carry printed or embossed identification on their front, rear, and spine, although this is an expensive operation. Modern binders have clear plastic sleeves on these faces permitting elaborate, but inexpensive, colored single-sheet company and equipment identification information to be inserted (Figure 18-4).

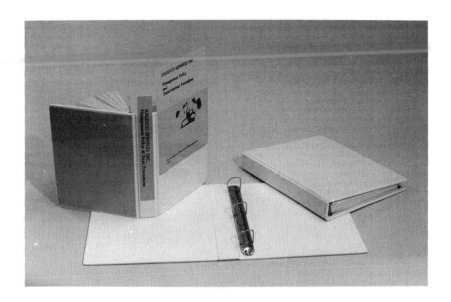

FIGURE 18-3
D-Ring Binder with Pockets

FIGURE 18-4
Typical Loose-Leaf Binders

Spiral and Plastic Binding: This type of binding involves the punching of both the covers and the sheets with a row of perforations at the back margin. For spiral binding, the holes are small and circular, through which a corkscrew-like wire is threaded (Figure 18-5).

For plastic binding (commonly known as *Cerlox*), the holes are rectangular and the claws of a long plastic clip hold the pages and covers in alignment. Both spiral and plastic binding can be performed as fully automated operations, or by small office-type machines on a one-at-at-time basis (Figure 18-6). Because of the cost of the materials, these binding methods can be considerably more expensive than the more common *perfect* binding when performed as automated high volume operations.

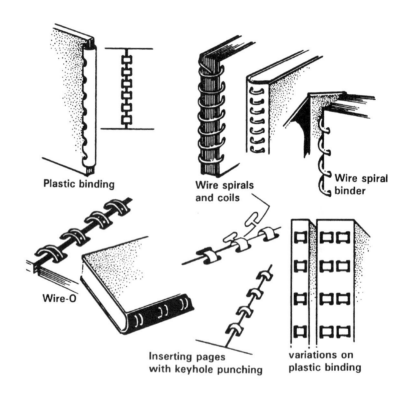

Plastic binding

Wire spirals and coils

Wire spiral binder

Wire-O

Inserting pages with keyhole punching

variations on plastic binding

FIGURE 18-5
Spiral and Plastic Binding

For the loose-leaf, spiral, and plastic binding styles, an allowance must be made in the inner margin (gutter) of the book for the punched holes.

Thermal Binding: This recently introduced equipment is an economical, professional quality binding system for technical manuals and books. An individual book, up to 3 inches thick, can be bound with a library-quality leather-like hard cover in a few seconds. The pages should be either single or a single fold, although folded drawings and the like, are acceptable as long as the edge of the inside fold does not come all the way to the spine.

The basic equipment comprises an electrically operated thermal gluing machine and pre-glued covers to accommodate books up to 3 inches thick and 15 inches in length.

FIGURE 18-6
Plastic Binding Machine

The covers are available in many materials from a leather-like material that can be embossed with gold printing to basic recycled board stock suitable for offset printing. Alternatively, a clear acetate cover can be used to display a cover page printed on light or medium weight paper, of any color.

It is available in office-size and commercial-size models and produces permanently bound books with incredibly simple ease. The only drawback for technical manual use is that, as with sewn binding, it cannot be amended by the replacement page method, although it is ideally suited for small manuals for such equipment as personal computers and printers and for software program instruction books. Figure 18-7 shows a typical thermal binding machine, capable of handling up to 10 manuals at a time.

BOOKBINDING

Generally, collating and wire stitching of ordinary booklets and brochures are comparatively simple and inexpensive. The operations necessary to obtain more permanent types of hard covers or case bound books are many.

FIGURE 18-7
A Thermal Binding
Machine
(Courtesy Southwest Binding Systems)

Preparations for bookbinding begin with design layouts. The task of keeping pages in order extends throughout every stage of preparatory work and printing. Each printed sheet, holding 8, 16, 24, or more pages must be folded to make up the sequence of pages called a signature. When placed in order, the signatures make up the entire book.

From this point on, two general bookbinding methods may be utilized; adhesive binding, which is known as *perfect binding* (not to be confused with *thermal binding*, previously discussed), or *sewn binding*.

Perfect Binding: (See Figure 18-8) This technique requires that all signatures making up the book be clamped together (1). The folds on the spine are then trimmed off, leaving an edge of perfectly aligned individual pages (2). Adhesive is then applied along this edge. If reinforcing is needed, a strong mesh cloth is applied to the spine and the adhesive penetrates both the cloth and the edges of each page (3). Covers are usually wrapped around the spine where the adhesive will also grip it.

Perfect binding can be done manually, partially by hand, or completely automatically. Normally, this style of binding is not considered as strong as sewn binding, but because perfect binding is relatively simple and highly mechanized, it has grown to be used extensively for large volume publishing and many publishers use it with hardbound covers.

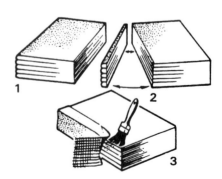

FIGURE 18-8
Perfect Binding

Sewn Binding: This is often called *case binding* because the cover encases the book. In sewn binding, the signatures are hand or machine sewn to one another. The thread is often looped around several hinge cords or passed through tapes, which cross the spine of the book (Figure 18-9). The tapes are normally made slightly larger than the width of the back of the book so that, later on when the hard cover is applied, the cloth stubs may be glued to the cover for extra strength and then be hidden by flyleaves, which are pasted over them.

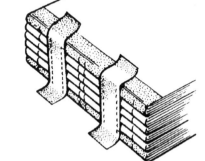

FIGURE 18-9
Hand-Sewn Binding

Many books are given a rounded spine which enhances their appearance and prevents the middle signatures from sticking out. The book is trimmed and adhesive is applied to the spine, penetrating between the sewn signatures. The rounded shape is formed by manual hammering or by a machine. The setting of glue helps to retain the shape. Loosely woven cloth or paper strips, called hinges, are then glued to the spine and the book is clamped firmly between boards to which these hinges are attached.

The outer cover material is placed over the assembly, formed over the edges and the flyleaves are applied. Finally, the covered book is clamped while the adhesives dry. In finishing, creases are made in the cover to make the book easy to open. Tooling, stamping, or embossing may be used to finish the cover.

FINISHING

Finishing is a general term that includes a number of different operations and specialties. Two finishing operations quite often used are embossing and die-cutting. It is usually applied to the front cover of a manual that uses soft vinyl or leather on the front face of the manual cover. The image is molded in embossing so that it is raised in relief. Molding is achieved by pressing the material to be embossed between brass dies on a press. This type of high quality bookbinding is rarely found on technical manuals.

TABBED INDEX DIVIDERS

The sections within technical manuals are often separated and indexed by the use of tabbed divider pages. There are many variations available, the use of these depending on such factors as client requirements, cost, and purpose served.

Generally, tabbed index dividers are constructed of heavy-grade card stock with the divider tabs cut into sets (commonly known as a *bank*). For full-size technical manuals, these banks can be as few as three tabs and as many as eight per page length (i.e., the number of tabs visible down the edge from the first page). The number of tab banks is directly related to the number of tabs required for the manual (such that there is an even distribution) and the amount of printing (if any) on each tab. Several sets of tabs may be needed; for example, for 20 tabs, one could use four sets of a five-tab bank, or five sets of a four-tab bank.

Printing on the tabs can be positioned in any plane, can be of any color, and can be on both faces of the tab. The body of the tab sheet can be printed with the title of the section/chapter or with other information as appropriate.

The tab and the perforated edge (if using three-ring binders) can be ordered with reinforcing Mylar facing. Generally, clear Mylar is used on the binding edge and colored Mylar on the actual tab. Both these additions ensure that a divider will stand the wear and tear of daily use.

When using index dividers, order them unperforated if the manual is to be collated, assembled, and perforated in one operation. This eliminates any chance of misalignment from different punching operations.

SHRINK PACKAGING

This operation has been used in recent years to replace brown paper packaging of printed material. As well as having the advantage of making the enclosed material visible through its wrapping, it also has advantages of increased production and reduced labor cost. The product is machine wrapped in polyethylene film, which is then tightly heat-sealed around it. With proper equipment and film, almost any kind of shape of product can be shrink-packaged.

CHAPTER 19

THE TECHNICAL EDITOR

INTRODUCTION

The first questions asked are: What is editing and what does an editor do? The definition of the term **edit**, as listed in the *Concise Oxford Dictionary*, is:

> *To prepare an edition of (another's work);*
> *set in order for publication (material chiefly*
> *provided by others); take extracts from and*
> *collate (films, tape recordings, etc.) to form*
> *a unified sequence; act as editor of (news-*
> *paper, etc.); reword for a purpose.*

Often in the absence of a professional technical editor, it often falls to the person coordinating the manual's production to act in this capacity. What can an editor do for a manual? Sometimes she is just there to clean up the rough draft so that the typist can read it. However, her responsibilities should not be so prescribed. A published document, be it a multivolume manual or a small report, is a company's face to the world, and on it may hinge a multi-million-dollar contract, or even the life of the equipment operator.

Whatever the case, someone is going to read it and base a decision, at least in part, on the intelligence that it conveys. Sweeping aside all the glossy presentation that is often part of a publication, it comes down to how well the facts are communicated from the originator's mind to the mind of the receiver. An editor can be a help or a hindrance to this communication; in an unfortunate number of cases she is often the latter.

In the business world where artistic merit is measured in profit, an editor's worth must be readily apparent to management, as is that of any other skilled professional. Industry is accustomed to the concept that you get what you pay for and what you get can be measured. An editor's product, the properly edited published document, is as valuable as, and parallel with, preventative maintenance, final inspection, value engineering, quality control, or any number of operations where the product is a piece of deliverable hardware.

In many areas the disciplines are so complex and sophisticated that not even *knowing the language* is going to ensure that the editor can grasp the direction of the discourse and do a credible job of editing. Many companies require their editors to have advanced degrees in a particular area of specialization — of necessity. The fields involve such a cascading of inter-related knowledge that a non-technically-schooled editor cannot pick up the depth of understanding necessary to work comfortably in them.

The editor's primary job, however, is still to ensure clarity, simplicity, and consistency in the presentation, no matter how technically adept she may be.

Given the editor has a strong academic background in English, journalism, or one of the other communications fields, what personal characteristics are desirable? The particular aptitudes required of a technical editor are the following:

a. An understanding of scientific and technical subjects (i.e., mechanical, electronic, medical, chemical, pharmaceutical, agricultural, and other disciplines).

b. An open and questioning mind.

c. Diplomacy.

Interest in scientific and technical subjects is an obvious attribute. This does not mean that the person must be skilled in any one of these subjects but rather that the person have some sympathy for the subject.

The second aptitude of an open and inquiring mind infers that although the technical editor may not be familiar with the topic at hand, she will learn quickly and will adopt the language of the subject from beginning to end.

Most companies have standard operating procedures that clearly define the levels of publication quality. If they don't they should (see Chapter 8 on the preparation of a formal manual specification). These quality levels are largely based on audience: an internal memorandum directed to a particular working unit would not have the same breadth of audience as the assembly instructions for a garden shed. Consequently, the need for constructive editing of published material may be an internal requirement based on the standards of the organization; it may be a necessity of the marketing environment of the company's product line, or it may be a customer's requirement.

Each organization has its own idea of how technical publications should be produced, and some companies may have several different areas with different products and different modes of operation. Inevitably, however, there is a person who edits the copy for grammar, accuracy, clarity, consistency, format, conformance to policy, and a multitude of other details. This person is the focal point and a prime mover in the publication process. One of the editor's most outstanding characteristics is empathy, the ability to feel and react to the written words in concert with the intended audience of the published document.

THE ROLE OF THE TECHNICAL EDITOR

There are many levels and kinds of editors. In some companies, an editor might need a Ph.D in nuclear physics in order to perform properly; in others, the secretary corrects the spelling and pays attention to minor grammatical oversights and that is good enough. Ideally, a technical editor will have an appreciation of communication and language, a strong background in English, a technical education, and training and/or experience in the company's field of interest.

The editor's responsibility deals with the communication aspect of the document rather than the technical aspects. Consequently, the editor should be strongest in communication skills, and at least conversant in the engineering principles involved in the subject. The editor should be able to ask intelligent questions of the technical author but not try to out-engineer the author in editing the copy.

Any technical author will react negatively to a fault-finding and pedantic attitude from an editor. On the other hand, almost every author is responsive to an interested and helpful technical editor who wants to work with the author to improve the product.

As a generalization, you might be working with three categories of technical authors. First, your company may employ professional engineering writers whose job it is to generate the technical copy for the publications. You should have no problem with these people; if they are professional writers, their copy will need very little editing and they will accept your stylish changes with no argument.

Second, you may be faced with a technical person who has been assigned the task of writing a manual text or alternatively has something significant to present as a paper before a scientific or technical society. This one requires special handling, but nothing like the third type, the company executive who has been invited to give an address to a IEEE or ASME convention. Let us take a typical category-two type for discussion, realizing that there are always variables on either side.

For discussion then, consider an author holding a B.S. and an M.S. in electrical engineering, four years out of an engineering university, who has written sections for progress reports that passed through many hands before they finally finished up as two lines in Appendix C of some major document.

Now, however, this author has the prime responsibility for a study report—50 pages with graphs, photographs, and much confusion and nervousness. This person has never met you or any other editor before and is therefore naturally suspicious.

Your first meeting is a general orientation in which you determine the points that relate to the overall approach to the report—audience, purpose, and what it is trying to impart. You do not come unprepared, however. If, for example, the document is in answer to a contractual requirement, the contract will specify some ground rules. You should know them, and let the author know that you know them — not to impress the author but to give him or her the assurance that you are affording the document the consideration it deserves.

A discussion of the flow cycle as it relates to the deadline of the report is also appropriated; you both then know what the important items are in the cycle and so can budget your activities accordingly. Then go over the outline, noting the major breakdowns and subparagraphs. A good working outline is not only an essential tool to constructive editing but also a means to a better understanding of the author's thought processes.

production personnel (or outside contractor if this is how your work is done), who must be instructed on the precise differences between this and the standard methods of production usually followed.

The easiest way to control the production and technical illustration efforts is to have a strong well-understood standard, with variations from it being the exception. Your standard should be part of your style guide: — for example, an appendix for line art and another for layout and production.

For a list of good style manuals, useful for creating your own version or as a reference, refer to the bibliography in this book.

EDITING PROCEDURES

It is essential that your editing be as neat, clear, and positive as you can make it: the word-processing operator will appreciate it and so will the author. There are several ways to approach the actual marking of the manuscript. Most style manuals and technical writing handbooks provide a comprehensive list of proofreader's markings, which may differ slightly from one another. A sample set is contained in Table 19-1.

However, unless both you and your typist are completely familiar with the exact meaning of each symbol, their use becomes more of a hindrance than an aid! Use only the most common markings, those that are relatively easy to understand and remember, and, if necessary, write out the instruction in the manuscript margin. Remember that indecisive, unintelligible editorial marks show indecisive thinking. If you do not know, find out! Schedule, cost, and audience determine the depth of editing, as long as clarity is not sacrificed to expediency.

PRINCIPLES OF EDITING

Anyone with a good basic knowledge of English can edit a document for grammar, spelling, and punctuation. These elements are important, of course, and they are quite obvious when they are incorrect. However, the additive factor of constructive editing is a far more demanding requirement of the technical editor.

If the editor is to contribute anything more than a simple style and mechanical edit, he or she must know more about the material than appears in the rough draft. To whom is the author talking? What is the technical background of the audience?

TABLE 19-1
Proofreader's Marks

MARKING	MEANING
wf //	Wrong font (size or type style)
☰ *Caps*	SET IN capitals
lc	Set words in LOWERCASE
lc & Uc	Lowercase and Initial Caps
rom	Set in roman (or regular) type
ital	Set in italic (or oblique) type
bf.	Set in boldface type
▽	Set superior character[66]
∧ 2	Set inferior character H_2O
[or]	[Move to the left OR move to the right]
‖‖	Align vertically
⊐⊏	Center
tr.	Transpose enclosed in ring matter
e	Delete, take out
stet	Let it stand (all matter above dots)
no ¶	No paragraph Run on
¶	New paragraph
#	Insert space (or more space)
⌒	Close up entirely; remove space
⌣	Less space between words
sp	Spell out (21 gr)
⌐	Break line or word
⊙	Insert a period ∧
⟨;⟩	Insert a semicolon ∧
⟨:⟩	Insert a comma then a colon ∧
|–|	Insert an EN dash ∧
=/	Insert hyphen ∧
(/)	Add parentheses ∧
[/]	Add brackets ∧

Each scientific discipline has its own idiosyncrasies, some of them important enough to be accepted when the audience will be within that field. For example, the bibliographical style for papers published in medical journals is considerably different from the style used in chemical engineering journals. Lexicons or technical dictionaries of disciplines are readily available.

First you must control the draft manuscript. Number the pages if this has not been done and, if you are handling other similar work, ensure that each document is identified as to its origin. It is embarrassing to find 37 pages of a manual chapter in the appendix of a sales document!

Remove any original artwork, photocopy it, and replace the copy in your master package of manuscript; then send the artwork on to the illustration department. Note on each copy of the artwork where it has gone and, if necessary, any details pertaining to its preparation. Even if very large drawings are involved, a photocopy of the title block will provide a ready reference and location. Artwork preparation is sometimes a lengthy process, so the editor must gain as much lead time as possible.

The first run through the text is an orientation tour, which will provide you with an understanding of the makeup of the document, and because no editor reading a draft copy can resist making marks, look for obvious grammatical problems and begin shaping the format. Mark the paragraph headings in order of importance and insert numbering. Take care to maintain consistency in this application. The occurrence of different numbering systems within the draft manual is quite common, especially with many contributing authors.

Indicate figure and table callouts, references, and footnotes so that they can easily be checked later. Check off the elements on the outline as you pass by them, making sure that they still make sense. Primarily, this pass is to get the sense of the manuscript, its development, what it is trying to achieve, and how effectively it accomplishes its objectives. If you locate major gaps in the document, now is the time to fix them.

We have already discussed the concept of constructive editing. You, as the editor, are the first outsider to read the manuscript and have none of the preconceived notions the author had when he or she began writing the material; you are able to look at it objectively, without the blind spots that pride of authorship gives. You are the intellectual and technological mean of the intended audience.

The complete and detailed editing procedure, which is next in the logical sequence, is the most difficult and exhausting. In previous operations, intelligence and common sense could often carry you through tolerably well. Now you must actively search for errors and inconsistencies that are there if you look hard enough. Disagreement between subject and verb, improper punctuation, misspellings, and *non sequiturs* of structure and logic are the most common offenses.

Again, the constructive editor will turn up inconsistencies and gaps that were passed in material the author may have scrutinized thoroughly and vigorously. The editor will actively seek these flaws and mark passages not understand for later clarification with the author.

In the preparation of a document for publication, your objective and subjective processes are so intricately linked that you may not be aware that elements of exasperation, evasiveness, or egotism have been allowed to distort the meaning. The process of "backing away" from the editing process is as difficult as it is necessary and it is an important function of the editor. It is a matter of obtaining objectivity and emotional distance—of seeing the words through the eyes of the reader. Put aside the document for a couple of hours, or even days if possible, while you do something entirely different.

Discount your previous impression of the material; look at it as though for the first time and as a complete stranger as you thoroughly review it again.

Remove Surplus Material: You want to deliver as simple and as functional a document as you can, for the purpose it is intended. Do not overburden the document by trying to make it do everything for everyone. If only a few readers will use the detailed information, or is it to be used only occasionally, include it in an appendix to eliminate disorder in the body of the document.

Clean It Up: Examine the text to ensure that it progresses in a clean direct line. If a word or paragraph does not contribute, eliminate it. If a sentence seems complex, determine whether it must be so because the subject itself is complex. Be suspicious of filigree—a certain amount of embellishment can freshen a section and is perhaps part of the author's style. Be certain that you are able to distinguish filigree from pretentiousness. The average draft manual contains a surprising amount of debris—perhaps one-fifth—so eliminate it. Do the document a favor by showing it off at its best.

Tune It Up: Now you are ready to see that the whole assembly works smoothly and logically. The traditional passages linking the components contribute purpose and value and should function smoothly and easily. Things are beginning to shape up. The pieces are falling into place. Final artwork is coming along from the illustrators and, along with photographs and other drawings, are fitted neatly into the slots left for them. This is the last chance to question whether it all works smoothly. Could a well-placed illustration clear up a muddy point?

Each organization has its own approval and production sequence for technical documentation. At some point, however, the editor and author must have a complete review of the material, examining in detail the changes and questions the editor brings up. Often the author, for technical or personal reasons, will find cause to object to changes that have been made. Some of these may be points of style or policy and will have to stand fast. Others will be revisions that can be negotiated to reach a compromise that is acceptable to both. In very few cases should it be necessary to quarrel over a few points on which mutual agreement cannot be reached.

Although the author gets credit or the blame, you cannot allow him to perpetrate an error. This author-and-editor meeting over the edited copy is one of the most critical tasks you have to face, and one that is often quite important to your personal success or failure in this undertaking. And there are other approval meetings for which you or the author, or both, may be responsible. These too can often be accomplished with greater ease, grace, and abandon by a tactful editor.

Final approvals, which may occasionally extend to the upper executive levels, quality checks by supervisory staff, and the ultimate finalization for camera or the printing press are areas in which you are fully involved, making or supervising the vernier corrections that are inevitable.

Now you send the camera-ready copy to the printer, with precise instructions on what is to be done. A sympathetic, responsive, and flexible printer can make your manual a thing of beauty — if she has the material to work with in the first place. High-quality copy input will permit her to give you a high-quality publication of which you both can be proud.

CHAPTER 20

A TECHNICAL HANDBOOK DEPARTMENT
From Concept to Operation

INTRODUCTION

A company making engineering products will find, as it begins to expand and obtain contracts for equipment of increasing complexity, that the need for technical literature increases in proportion. It will soon become apparent that technical sales literature—supplemented by copies of internal reports, specifications, and instructions from engineers, technical advice by fax, correspondence, or over the telephone, and working drawings—is quite inadequate for the user's needs. Only a technical handbook will meet these needs, and such a handbook is likely to form part of a contract.

To satisfy the initial demand, a full-time technical writer may be engaged, to relieve the engineers of the burden of producing a stop-gap publication. Continued expansion, however, will sooner or later require a self-contained department for the preparation of technical manuals, technical bulletins, modification leaflets, and so on.

This may develop gradually from the labors of the solitary writer, or, alternatively, the company may decide to appoint a suitable executive to establish a fully equipped technical publications department.

In this chapter, the problems of setting up and laying out such a department, and determining its relation with other departments inside and outside the company, are discussed.

ESTABLISHING THE DEPARTMENT

In creating any department, much depends on the quality of the executive appointed and on the support he receives from his superiors. In addition to the qualities common to all departmental heads, the executive should be fully experienced in the preparation and production of technical handbooks within the appropriate branch of engineering.

The department head should be able to edit technical matter, inspire the writers and illustrators to do their best work, and provide them with efficient administrative services. He should have a sound engineering training, a good knowledge of layout and printing, and an understanding of the requirements of any official bodies that have jurisdiction over the handbooks to be provided. The position is certainly no sinecure, to be awarded to a long-serving but inexperienced senior employee from another department within the company.

The department, when created, may be responsible to the senior engineer, the service manager (where there is one), or possibly the sales director. In some large companies, the technical publications department may be grouped with a technical library or photographic and training departments under some appropriate all-encompassing title and administered by a senior executive.

As a first step, the newly appointed head will need to know the probable load on the department during the first year of its operation. This may include arrears of outstanding work and certain new projects that are anticipated during that period. From this knowledge he can deduce how many experienced writers will be required initially. Assuming he decides that two writers will serve the department's needs, he can now estimate the amount of space they will occupy. Each will need basic office space and will collect a certain amount of material requiring storage space, in the form of one or two filing cabinets and a further cabinet for storage of artwork.

An experienced word-processing operator/secretary (preferably one accustomed to technical terminology) will be required. To give the typist time to organize the necessary filing systems, some of the draft typing will probably have to be assigned to some other word-processing group during the early stages.

One or two technical illustrators will be needed, and where there is a sufficient volume of halftone work to justify it, one of these artists should

Illustrators

Technical illustrators should be readily accessible to the technical writers, and vice versa. Where space is available, they should be accommodated in an area near the writers, not banished to some remote drawing department. The convenience of instant discussion on a subject is important.

Allocation of Files

Either a filing cabinet can be allotted to each writer or a central filing system can be established, divided according to projects. The decision will be influenced mainly by the nature of the projects. Where a writer is engaged on a lengthy handbook that will occupy him fully, he will naturally prefer his own filing system. On the other hand, if many small items have to be covered and writers are likely to switch frequently from one book to another, a communal system in which each writer can find his way unaided is probably better.

FURNITURE

Large work tables should be provided. Even in the age of computer-driven illustration, there is a need for areas on which artwork can be spread for reference and marking. Otherwise, there is the risk of damage to a piece of artwork that is occupying the whole surface of a desk while a writer is rummaging beneath it for pencils. The table will also be useful in collating internally printed publications.

Technical writers, involved with the preparation of illustrated parts lists should have sufficient wall space and fittings in their offices to hang the large E-size engineering drawings from which these lists are created. During the preparation of a complex IPB manual, the writer may have to refer to and store almost 100 drawings of varying sizes within his office.

Care should be taken to avoid locating equipment such as pen plotters, color printers, and photocopiers close to the writing staff. The operating and standby mode noise from these machines can be extremely distracting in a quiet environment. Placing these machines behind noise-cancelling barriers will greatly improve such situations.

CONTACT WITH INTERNAL GROUPS

During their work, the writers and illustrators may have dealings with some or all of the following groups: engineering departments, CAD/CAM sections, service, sales, publicity, photographic, contracts, patents, and specifications departments.

Engineering Departments

Test and development engineers are the source of much of the writer's information and also present one of the principal hurdles in the approval procedure. Their goodwill is vital to the success of a technical publication.

Illustrating and CAD/CAM Departments

Computer-designed and drawn working and information illustrations form one of the chief sources of pictorial information. Approved circuit diagrams, wiring layouts, installation outlines, general assembly drawings, and mechanical detail drawings are the basis of many of the illustrations used in manuals. Also, details such as fits and clearances, weights and dimensions, and materials and finishes are usually obtained from these departments.

These departments are often responsible for naming components and assemblies, composing labels, and allocating reference symbols. This can cause difficulties when writers disagree with the names applied, apparently arbitrarily, by draftsmen to items that have to be described in a handbook. Inconsistencies can often arise from draftsmen on similar, but unrelated, projects calling identical parts or services by different names. Confusion is also caused if the same name is given to several items of a general nature (such as *bracket* or *cover*), which have then to be distinguished in the handbook by qualifying adjectives added by the writer.

Compromises may have to be accepted in matters of this kind. It is better for an unsatisfactory name to be applied consistently to an item than for variations to creep in, with possible confusion to the reader. However, the head of the publications department should stand firm whenever there is a clear case of error or inconsistency in work originated by the artwork department.

Service Department

Where a service department exists, this will usually be the authority on policies of maintenance and repair procedures. Any literature that may be prepared by service engineers for publication should pass through the publications department for editing, layout, and printing, to ensure a uniform company standard of presentation.

Sales, Publicity, and Training Departments

Since technical handbooks are predominantly an after-sales provision, the sales department will be concerned with the availability and prestige value of a handbook. The book will help to inform sales engineers and executives, who will readily report any defects that they and the user may find in it. It may also form the basis for factory-training of a customer's staff.

The publicity department will often rely on the technical publications department for technical information on which to base their sales literature.

Photographic Department

In some companies, the photographic services are under the control of the publications department. Where this is not the case, the publications department may have to compete with others for their services. However, the advent of computer-driven photograph manipulation software programs and CD-ROM libraries of photographic material has given the illustrator significant control over the preparation of color and halftone photographs that go into manuals.

This is, of course, dependent on the company having personnel trained in the use of these programs, plus the availability of the equipment in-house. Such equipment and programs have reduced reliance on the traditional photographic departments for the provision of the photographic material for technical manuals.

Contract, Patents, and Specifications Departments

These departments will be concerned, respectively, with contractual requirements for handbooks, statements of patent rights, and agreement of performance and test figures, quoted in a handbook, with those appearing in company specifications. Any of these matters may require discussion with the publications department.

CONTACT WITH EXTERNAL GROUPS

These include departments within the customer's company, printing firms, material suppliers, industrial photographic companies, and temporary staffing agencies.

Client Relationships

Inquiries, complaints, and (less often) compliments from the client are likely to be routed directly to the publications department for attention.

Information is often exchanged by the publications department of the manufacturing company and interested departments of the client, especially when the former is a subcontractor to the latter. A helpful attitude toward the purchaser is all-important. There should be no specious excuses for failure to meet target dates or for the supply of incorrect information. Promises should not be made unless they are virtually certain to be kept, as far as can be judged by prevailing circumstances.

On the other hand, a distinction should be drawn between genuine complaints about defects or errors in a handbook and mere hair-splitting quibbles. Also, callers with queries on matters outside the scope of the publications department (such as points of design, provision of equipment, and doubts about the validity of information quoted in good faith in a handbook) should be passed over immediately to the department most directly concerned.

Printers and Suppliers

Many services may be needed that are not available within the company. How these needs are met will depend on the cost allocation and the degree of freedom allowed to the head of the publications department.

The meager basis on which some departments are required to work can make external purchasing a matter of great delicacy. In general, urgency costs money. Undue haggling with a printer about the cost of a job simply results in one being consigned to the end of the queue, the more lucrative clients taking precedence. To obtain exactly what is required in style and quality from a manufacturer/supplier of paper, binders, tab dividers, and so on, is often difficult. This difficulty is greatly increased if such purchases have to be made remotely through an internal purchasing department, multiplying the risks of error and omissions during routing of orders, and causing extra delay in delivery.

Whatever the ordering procedure, personal contact between the head of the publications department and a representative of the supplier is essential.

Contract and Temporary Staff

In an emergency, it may be necessary to hire word-processing typists, illustrators, and even technical writers from organizations specializing in these services. Staff so obtained will naturally be more expensive than the company's own staff. Apart from increased overhead costs, time will have to be spent in familiarizing the new staff with particular projects. Accordingly, it is much better to engage them at the beginning of a project, while there is time for them to settle in, than to wait until the company's own staff is overloaded.

TIME AND COST FACTORS

Timesheets or job cards, recording the time spent on individual projects by members of the staff, are often vital to compiling costs. When they are correctly compiled, these records can be analyzed to show the amount of time spent by different persons on a particular project.

Advance estimates of time and cost in the preparation and production of technical handbooks are sometimes necessary for factoring into a contractual bid. They are, however, liable to be upset by many factors such as design changes, diversion of effort to other projects, and misconceptions regarding the exact scope of a book. In all creative work, there is a natural working speed that is peculiar to the individual concerned.

It is usually not possible to reduce the preparation time by dividing a project planned by a particular writer and giving part of it to another. Even if the writers are in complete harmony, which is unlikely, the result will probably be disjointed and inconsistent, while if the equipment is complex the result may be chaotic. It is certainly never possible for a writer to increase his/her speed of working to make up for lost time caused, for example, by delay in the supply of information by other departments.

When a series of technical manuals is required for a range of similar items of equipment, it is possible, after completion of the first book, to budget fairly accurately for the subsequent ones.

Standard methods of layout and production, as far as the diversity of products and users' requirements will permit, and close collaboration between writers to prevent unnecessary duplication of writing effort will help to reduce the time and cost and make them more predictable.

Where the cost of a technical manual is not included in the contract for the equipment, it is not usually practical to recover the full cost of the manual from the purchaser directly. After a nominal charge has been made, the difference must be obtained either from overhead costs or from an allocation made to the publications department. This depends on company policy. Included in any estimate for the supply of a technical manual should be a realistic allowance for future amendments and reprinting.

When the contract for the manufacture and supply of equipment is awarded, a clause may be included requiring that suitable technical information is to be provided, in the form of a technical manual. The cost of providing this technical manual is then wholly incorporated in the contract price. If there is also a penalty clause for failure to deliver the equipment by a specified date, the technical manual will be covered in this clause. The completion date for the technical manual will thus be the same as that for the equipment itself.

In attempting to reduce costs, "economies" are sometimes made by under-staffing the department and lowering the standards of presentation to the minimum considered acceptable to the reader. Apart from the poor impression this is certain to make on the purchaser, such false economy will actually increase costs by alienating good (and therefore efficient) staff and encouraging makeshift methods of preparation and production. This also increases the likelihood of errors, causing unnecessary delay, requiring costly remedial action, and bringing the department into disrepute.

The only true economy is achieved by selecting the staff with care, maintaining an efficient administrative system, and giving everyone a proper pride in their work by setting the highest possible standards. This will ensure that the best results are obtained within the limits of the processes available. Then the technical manual, instead of being a dismal stopgap, will enhance the reputation of the company and help to sell more equipment.

CAPITALIZATION RULES

INTRODUCTION

In the German language, the first letter of every noun is capitalized, a fact that may seem bizarre to users of modern English. But German is not the only language to have followed that custom. In the latter part of the nineteenth century, some English writers such as Thomas Carlyle followed this pattern.

Today, however, the use of capital letters in the English language is not so predominant. Rules of usage have greatly reduced the field of application, and modern writing techniques have removed much of the unnecessary formality. Although careful writers exercise selectivity even within the presently restricted field of use, some confusion can still exist. This appendix is included as a reference section to illustrate some basic rules of usage.

REPRESENTATIVE EXAMPLES

Some representative examples of names and terms where uncertainty sometimes arises are presented below:

James Walker, mayor of New York; Mayor Walker; the mayor of New York; the mayor.

Alan Hale, president of Ontario Hydro Corporation; the president of the corporation.

Dade County; the county.

the Province of Ontario; the province.

Bruce Mines Township; the town of Bruce Mines.

the Humber River; the Humber and Credit rivers.

Lake Ontario; Lakes Ontario and Erie.

the Toronto Board of Education; the board.

the Committee of Foreign Affairs; the committee.

the federal government.

The Bureau of the Census; the bureau.

city council, but Virginia Beach City Council.

The Hudson's Bay Company; the company

Environmental Assessment Act; the Act.

uranium 235; U-235.

central Asia, but Southeast Asia.

FORMAL CAPITALIZATION RULES

Examples of formal capitalization are given in the following sections.

Initial Word of Sentence

Capitalize the first word in a sentence, as is done in this sentence.

Quoted Material

Capitalize the first word of a direct quotation when the latter is separated from the introductory statement.

> The chief engineer replied, "We have had only one equipment failure in the last 10,000 hours."

Secondary Statements

Capitalize the first word after a colon if it begins a complete statement.

In summary: The evidence has proved a clear case of metal fatigue.

My final point: Keeping abreast of advances in this technology will be of immense benefit to the whole department.

Displayed Material

Capitalize the first word in displayed material designated by an identifying letter or number (i.e., in subparagraphs).

Methods of capacitance measurement include the following:

1. Bridge method

    (a) Schering bridge

    (b) Wien bridge

2. Capacitance-meter method

    (a) Dynamometer type

    (b) Voltmeter type

3. Resonance method

    (a) Grid-dip oscillator

    (b) Q meter

Titles of Works, Captions and Headings

Capitalize the first word of each principal word in a title or heading. Use italics if a title of a publication.

The Chicago Manual of Style

Research and Design

Measurement of the Second Harmonic

The Loss Factor

Reinforced Concrete Handbook

Peter Norton's Advanced DOS 5

Capitalize both the full names and the shortened names of government agencies, bureaus, departments, services, and so on.

National Film Board

Faculty of Law

Statistics Canada

Bureau of Pension Advocates

Ombudsman of B.C.

Law Reform Commission

Do not capitalize the words government, administration, federal, and so on except when part of the title of a specific entity.

The U.S. Government is the largest employer in the state.

She hopes to work for the federal government.

Capitalize departments or divisions of a company.

Administrative Services Department

Design and Development Division

Capitalize ethnic groups, factions, alliances, and political parties, but not the word *party* itself.

He spoke for the Chinese community.

The Republican party held its pre-convention in July.

North Atlantic Treaty Organization; NATO

The Communist bloc vetoed the proposal.

NOTE
Political groupings other than
parties are usually lowercased

He represents the left wing of the Canadian Labour Congress,
but the Right, the Left

Negro and Caucasian are always capitalized, but blacks, whites, or slang
words for the races are lowercased.

Capitalize scientific names for genus, but not species.

Drosophila melanogaster (abbreviated *D. melanogaster*)

Homo sapiens

Do not capitalize the word *the* in the reported title of a newspaper or
magazine, even when it is part of the title.

the *New York Times*   the *Journal of Metallurgy*

Proper Noun

Capitalize each proper noun.

Princess	Senate (the)
Father (religious)	French
Riel Rebellion	Ontario Hydro
North York	Yonge Street
Crown (the)	Group of Seven
January	Brazil
Boy Scout	New York Philharmonic
Peace Corps	Nobel Peace Prize
Holy Ghost	Christmas Eve (holiday)

Do not capitalize names that have become common nouns (such as units of measurement): *henry, ohms, watt.*

Do not capitalize the names of seasons: *spring, summer, fall, winter.*

Abbreviation of Proper Nouns

Capitalize the abbreviation of a proper noun.

AECL	Ont.	N.W.T.	Mr.
Sept.	Ms.	Dr.	Ave.

Titles as Generic Terms

Do not capitalize a title that stands alone and describes a role rather than the person fulfilling it.

The director of personnel approves promotions.

The governor general represents the monarch.

Most articles are approved by the managing editor.

Adjectives Derived from Proper Nouns

Capitalize an adjective formed from a proper noun.

De Broglie equation	*Fourier* analysis
Laplace transformation	*Parseval's* theorem

Compound Noun or Adjective

In a compound noun or adjective, capitalize only that part which is a proper noun or adjective. If the compound starts a sentence, also capitalize its initial letter.

British thermal unit	inter-American
pre-Darwinian	post-Cambrian
trans-Siberian	pro-Asian

In the abbreviation of a compound, capitalize only the letter standing for the proper noun or adjective, for example, **Btu** (British thermal unit), dBa (adjusted decibel).

Emphatic Element

Use full capitals occasionally to emphasize one or more words (more emphatic than using the "quotes" form).

Reset the switch to the STANDBY position.

DANGER, RADIATION HAZARD.

These fumes are **VERY POISONOUS** and personnel must use personal respirator equipment!

Before covering, CHECK THE PIPING FOR LEAKS.

Salutation of Letter

Capitalize the salutation of a letter.

Dear Mr. Wright:	Sir:
Dear Sir:	Dear Madam:
Gentlemen:	Dear Mr. President:

Do not capitalize the word *dear* when it is not the first word in the salutation.

| My dear Sir: | My dear Mr. Clark: |
| My dear Madam: | My dear Ladies: |

Complimentary Close of Letter

Capitalize the first word in the complimentary close of a letter.

| Respectfully yours, | Yours, |

Sincerely,   Yours sincerely,

Very truly yours,   Yours truly,

Title of Respect

Mister   Madam

Miss   Mesdames

Gentlemen   Sir

Abbreviation of a Title of Respect

Capitalize the abbreviation of a title of respect.

Mr.   Messrs.   Mrs.   Mne

Professional and Personal Titles

Personal titles are capitalized only if they precede the name and are *not* separated by a comma.

Professor Colin Menzies

the finance minister, Richard Quayle

Capitalization is optional if the title follows the noun.

Bruce Roberts, president of the corporation, or

Bruce Roberts, President of the corporation

Capitalize the abbreviation of a professional title used with the holder's name.

Capt. Jan Chislin   Lieut. Comdr. Val Macri

Maurice Hagus P.Eng   Prof. Colin Stagg

Sen. John Rowe   Maj. Peter Evans (Ret.)

Names of Academic Degrees

Capitalize the principal words in the name of an academic degree.

Associate in Arts   Master of Arts

Bachelor of Science   Mechanical Engineer

Doctor of Dental Surgery   Pharmaceutical Chemist

Abbreviation of Academic Degrees

Capitalize the abbreviation of an academic degree.

B.A.   LL.B.   M.S.   A.C.S.   Ph.D.

Litt.D.   B. Arch. E.   M.S. in L.S.

Trade Names

Capitalize (and italicize) a proprietary trade name, but do not abuse these names, for example, "areas of the kitchen were finished with a butcher block patterned *Formica*" or "... were finished with a butcher block patterned plastic laminate" are correct. Do not write "personnel can xerox copies of the forms to...." Xerox is the name of a company and the brand name of one photocopier. It is not a synonym for the word photocopy.

Some current trade names are often confused with generic products or processes. This list is of some of the more familiar names:

Fiberglas (insulation)   *Pyrex* (dishes)

Kodak (film)   *Kleenex* (tissues)

Powerstat (equipment)   *Frigidaire* (refrigerator)

Plexiglas (plastic)   *Vaseline* (petroleum jelly)

Polaroid (camera)   *Band-Aid* (dressing)

Formica (material)   *Orlon* (material)

To overcome the tendency to use trade names directly, write the first example as "pack with glass fiber insulation to the level...," rather than the less preferable "pack with *Fiberglas* to the level...." The latter form tends to indicate a preference for a particular brand rather than a general requirement.

Acronym

Use full capitals for some acronyms.

   CANDU (Canadian Deuterium Uranium)

   QANTAS (Queensland and Northern Territory Air Service)

   DOS   (Disk Operating System)

Points of a Compass

Capitalize the names of specific geographical areas.

Far West	Pacific Northwest
Middle East	South America

Do not capitalize words that simply indicate direction.

western provinces	the east coast
east of the Rockies	directly to the south

Some regional terms, such as *Prairie Provinces*, seem to be either part of a "precise descriptive title" or "merely suggest position," depending on your viewpoint. Since authorities can be found on both sides of this gray area, choose whichever you are more comfortable with, and capitalize accordingly.

Compound Numbers

At the beginning of a sentence or in displayed material, capitalize the first word in a written compound number.

Thirty-three	One-half	Five sixty-fourths
Sixty-nine	Seven-eighths	One hundred ten

MATHEMATICAL AND SCIENTIFIC TERMINOLOGY

INTRODUCTION

Since accurate typographical presentation is crucial to understanding equations and scientific terminology, technical writers and editors should read this section of the manual with special care: its aim is to provide information about the limits of existing compositional resources and to indicate standard methods of layout that will ensure correct and efficient presentation.

AVAILABILITY OF CHARACTERS FOR TYPESETTING

The Latin and Greek alphabets are readily available for use in technical material. The Latin alphabet is available in six type styles, or *fonts*: **roman**, in standard (light and upright), **italic** (light and oblique or slanted to the right), and **Univers/Helvetica** (sans serif) in both upright and italic. Both these fonts in their standard or italic form can be set in the **boldface** (heavy) style.

Greek symbols, diacritics (accents), parentheses and brackets, mathematical signs of relation, mathematical operators, phonetic symbols, figure symbols, and so on are also generally available. Any of these symbols may be used if necessary, but mathematical notation should be kept as simple as possible. Some limitations exist due to the types of software programs being used; however, these shortcomings (if any) would be indicated to the technical writer by the editor concerned.

A mathematical expression used repeatedly throughout a paper should be defined in terms of an appropriate symbol the first time. This symbol can then be used as shorthand for the expression throughout the document. Look for opportunities to use this technique, which will reduce the chance of errors in typesetting.

MARKING OF MATERIAL

The importance of legible mathematical material cannot be over-emphasized. Since word-processed text is far more legible than handwritten material, request that the mathematical material prepared by authors be typewritten as much as possible. Handwritten material, if submitted, must be neatly lettered.

CLASSIFICATION OF MATERIAL

Many handwritten letters, numbers, and special symbols look alike and are hard to identify. Each should be identified the first time it occurs in the draft document, and thereafter if any ambiguity is still possible. Write the identification in black pencil above the symbol or in the margin.

Typed material has the advantages of legibility and consistent formation of character, but some confusion is possible. On most printers, the letter l and the number 1 are the same, and the writer must distinguish between them, although modern word-processing systems are sophisticated enough to overcome this problem. Similar precautions should be taken to distinguish between a capital O and a zero (0) and between a capital X and a multiplication sign ($\times$).

MATHEMATICAL TERMINOLOGY

Roman Versus Italic Type

According to conventional typesetting practice, Latin letters used as mathematical symbols are set in *italic* type to distinguish them from ordinary roman text. The document text will therefore have all unmarked Latin letters, that are obviously not words, set in italic type; for example, the statement

> For $2x_r e^x = 0$, we obtain …

would be set as

> For $2x_r e^x = 0$, we obtain …

Authors should mark a letter for italic type (by underscoring in black pencil) only if the letter might be mistaken for a word; for example,

> When $\underline{x}$ is a number dependent on …

will be set as

> When x is a number dependent on …

Since subscript words need to be set in roman type, authors should mark such words in pencil only when they might be mistaken for symbols; for example,

> The values of V/in\and V/out\both decrease with…

clearly should be set as

> The values of V_{in} and V_{out} both decrease with …

Some Latin letters, considered abbreviations of words, are properly roman instead of italic, for example, chemical symbols (CO, Ne), most multiletter abbreviations (fcc, ESR, exp, sin), and most units of measure (k, Hz). But the editors are trained to spot these, and authors need not mark them for roman type unless confusion is especially likely; for example,

$$\rho^2{}_s \qquad\qquad {}^{238}U \qquad\qquad {}^{18}O \qquad\qquad {}^{14}C$$

The staggering in the first example is due to the inability of the current software to set the subscript and the superscript on the same vertical alignment.

Displayed Equations

Equations should be set on separate lines below the text. Where equations are longer than the width of the text column, certain steps are necessary to present these.

In its original form the following equation was far too wide for this column and would have had to be set as follows:

Total Current at Maximum Voltage

$$= \frac{\text{Total Current} \times \text{Rated Voltage}}{\text{Maximum Voltage}} \qquad (1)$$

This is an unwieldy wordy equation that does not look or read well. The author should consider the alternative form shown below:

$$I_{\text{Max}} = \frac{I_{\text{Tot}} \times V_{\text{Rated}}}{V_{\text{Max}}} \qquad (2)$$

where

$$I_{\text{Max}} = \text{Total current at the maximum voltage}$$

$$I_{\text{Tot}} = \text{Total current}$$

$$V_{\text{Rated}} = \text{Rated voltage}$$

$$V_{\text{Max}} = \text{Maximum voltage}$$

Although this layout takes up more space vertically, it is easier to read and provides a clearer, precise equation.

Equation Numbers

Only displayed equations should be numbered.

For equation numbers in the body of the document, use arabic numerals in parentheses: (1), (2), (3), and so on. Number all equations consecutively throughout the text. Make reference to equations with parentheses intact, that is, (6), (7), and (8), *not* (6, 7, 8).

Place equation numbers flush with the right margin, as shown above. Leave a space at least the width of two characters between an equation and its number.

Fractions and Negative Exponents

A fraction can generally be represented in three different ways:

with a fraction line:

$$= \frac{a + b}{c} \tag{3}$$

slashed with a solidus:

$$= (a + b)/c, \tag{4}$$

or with negative exponents:

$$= (a + b)c^{-1} \tag{5}$$

Do not mix built-up and slashed forms unnecessarily within one equation.

Write it as

$$\frac{a}{b} = \frac{B(E_0) + c}{f_1 + f_2} f(\omega) \tag{6}$$

instead of

$$a/b = \frac{B(E_0) + c}{f_1 + f_2} f(\omega) \tag{7}$$

However, a built-up fraction within the numerator or denominator of another built-up fraction is ungainly and hard to typeset.

Write it as

$$B' = \frac{[3J/(J+1)]T_N\, m^2}{En - Bn^2b^2n^2T_N} \tag{8}$$

Any rational fraction can be put on one line in the special upright form, for example, ¾. This form is awkward to set in subscript and superscript form, so use a slashed fraction in these situations; for example,

$$I_{sym(rms)} = \sqrt{2}[1 + e^{-(\omega/15)}] \tag{9}$$

Elsewhere, use the upright form instead of forms with a solidus as often as possible. Write ¼x instead of $x/4$, $(1/4)x$, or $1/4\ x$; never write $1/4x$ unless you mean $1/(4x)$.

Write

$$\sin(¼x) \quad \text{and} \quad ¼\sin x$$

instead of

$$\sin(x/4) \quad \text{and} \quad (\sin x)/4$$

Never write $\sin x/3$.

When using the solidus, ensure that your meaning is clear and unambiguous:

$$a/b + c \quad \text{means} \quad (a/b) + c$$

Use parentheses, or write $ab^{-1} + c$, to dispel any possible confusion.

Never write

$$a/b/c$$

Use the form

$$(a/b)/c \quad \text{or} \quad a/(b/c)$$

or better still,

$$ab^{-1}c^{-1}$$

Make sure that the fraction line clearly delimits the numerator and denominator of a built-up fraction; for example,

$$\text{does} \quad \sin\frac{a}{b} \quad \text{mean} \quad \frac{\sin a}{b} \quad \text{or} \quad \sin\left[\frac{a}{b}\right] \quad ? \tag{10}$$

Clearly state your requirements; do not expect the editor to resolve such ambiguities.

Parentheses

According to accepted convention, one works outward with parentheses and brackets, according to the scheme:

$$\{ \quad [\quad (\quad) \quad] \quad \}$$

Small parentheses and brackets are available for use in subscripts and superscripts. Oversize parentheses and brackets are available for use in displayed equations around expressions containing built-up fractions or integral, summation, or product signs [as shown in Equation (10)].

In general, too many parentheses are better than too few. But wholesale use of parentheses obscures rather than clarifies and, especially in displayed equations, wastes labor and space.

All of the parentheses in the following example are extraneous:

$$\left[\left(\frac{\Delta + 5}{a}\right) + \left(\frac{3x^3}{(0.06\pi/a)}\right)\right] \tag{11}$$

Mathematical Functions Set in Roman Type

Standard abbreviations for functions are set in roman type; however, care should be taken, when using such functions, to respect the following conventions.

(a) A function is closed up with its argument:

$$\text{Tr}Q \qquad \tan(\eta^2/\mu)$$

(b) The function of a product may be written without parentheses:

$$\sin xa \quad \text{means} \quad \sin(xa)$$

However, put parentheses around the product whenever there is a chance of confusion:

$$\sin(x^2 a^{3/2}) \qquad \sin(\tfrac{3}{4}x)$$

Note, nevertheless, that an argument is assumed to stop as soon as another function appears:

$$\sin x \cos a \quad \text{means} \quad (\sin x) \cos a$$

(c) $\sin x/a$ is ambiguous. Write $(\sin x)/a$ or $\sin(x/a)$, depending on your meaning.

(d) An argument stops at a plus or minus sign:

$$\sin x + a \quad \text{means} \quad (\sin x) + a$$

(e) A plus or minus sign should never directly follow a function.

Write

$$\sin[-(x + a)] \quad \text{instead of} \quad \sin -(x + a)$$

Radicals and Bars Over Groups of Symbols

The "roof" of a radical or a bar cannot easily be set over expressions, as in

$$\sqrt{a}, \ \sqrt{a+b}, \ \text{and} \ \ \overline{a+b}$$

due to software complications. An exponent $^{1/2}$ or special brackets should be used instead, for example, $(a^2 + b^2)^{1/2}$. A bare "unroofed" radical may also be used, but always with a liberal use of parentheses to avoid ambiguity.

Write

$$\sqrt{(x^3 a)} \qquad \text{or} \quad (\sqrt{x^3})a$$

$$\sqrt{(x^3/a)} \qquad \text{or} \quad (\sqrt{x^3}/a)$$

$$\sqrt{(x^3 + a)} \ \text{or} \quad (\sqrt{x^3}) + a$$

Usage will depend on your meaning. Clearly, as substitutes for roofed radicals, forms with an exponent $^{1/2}$ are usually less troublesome and more readable than forms with a bare radical.

Dots

Dots over letters (e.g., $\dot{C}$, $\dot{I}$, $\ddot{O}$, etc.), while being relatively easy to set, do not reproduce very well except on very high quality documents such as textbooks and formal technical manuals. Authors should consider using a prime or some other more substantial symbol, for example, $\check{D}$, $\dot{g}$, or S', to indicate special terms.

SCIENTIFIC MATERIAL

Generally, the rules of the preceding section on mathematical form apply to this section also; however, there are some specialized areas that require attention.

Nuclides and Isotopes

Nuclides are atoms defined by their atomic (proton) number and mass (nucleon) number. Different nuclides with the same atomic number are called *isotopes* or *isotopic nuclides*. Different nuclides with the same mass number are known as *isobars* or *isobaric nuclides*.

A particular nuclide is specified using information set in superscript and subscript positions to the left and right of the element symbol.

The mass or nucleon number is set in superscript on the left, for example,

$$^{14}N \qquad\qquad ^{16}O$$

The atomic or proton number is set in subscript, also to the left, for example,

$$_{7}N$$

A particular nuclide is specified using information set in superscript and subscript positioned to the left and right of the element symbol.

The state of ionization is set in superscript on the right of the element symbol, for example,

Pb^{2+} Cl^-

The number of atoms in an entity is set in subscript, to the right of the element symbol, for example,

$^{14}N_2$ PO_4^{3-}

Note the inability of the software program to permit the alignment of the superscript and subscript in the last example. The form shown is acceptable for most scientific papers.

Similarly, in nuclear physics the number of neutrons in the nucleus, is set in subscript on the right, for example,

$^{235}U_{143}$

Authors may use an alternative form, that is, give the name of the element with the mass number following; for example, uranium with its mass number of 235 may be written as uranium-235. It is also sometimes written as U_{235} or U-235 but these methods are not recommended.

The presentation of chemical reactions presents no difficulty provided the author is quite explicit within the manuscript.

Some examples of equations indicate the possible layouts:

$$AgNO_3 + HCl \rightarrow AGCl\downarrow + HNO_3 \tag{1}$$

$$MnO_2(s) + 4HCl \rightarrow 2H_2O + MnCl_4 \tag{2}$$

$$MnO_2(s) + 4H^+ + 4Cl^- \rightarrow 2H_2O + Mn^{4+} + 4Cl^- \tag{3}$$

C

USING THE METRIC (SI) SYSTEM

INTRODUCTION

This appendix covers the basics of the International System of Units (SI) with particular reference to the rules of writing or typing the units and symbols. It has been prepared to assist writers and editors in the correct and uniform application of metric terminology likely to be found in technical manuals.

The information contained herein has been extracted from the *Metric System of Measurements (SI)* published by the U.S. Department of Commerce and the *Metric Practice Guide* from the Canadian Standards Association.

The correct use of the symbols and mathematical designations must be clearly understood by all personnel preparing documents for publication. The improper use of a letter or mathematical symbol could alter the entire meaning of an expression and lead to errors.

RULES OF USAGE

To achieve the basic intent of the SI measurement system, which is to improve communication on an international basis, the following set of rules for writing or typing has been established.

Rule 1: Symbols, with the exception of Greek character symbols, are expressed in upright type only, regardless of the style of type used within the rest of the text. Numbers are also always set in upright text, whether subscript or superscript.

Rule 2: Symbols shall not be followed by a period except when at the end of a sentence.

Correct Method:

25 kg   67 Mg   4.45 kN   60 Hz

Rule 3: Symbols are never expressed in the plural form.

Correct Method:

4600 N   638 km   32 GHz

Rule 4: Symbols shall not begin a sentence in any text.

Rule 5: In text, symbols are to be used when associated with a number. When no number is involved, the unit is to be spelled out.

Correct Method:

The area of the room is 250 m^2.

The area of the common room is to be measured in square meters.* (**NOT** expressed as m^2)

Rule 6: Word abbreviations shall not be attached to or used in conjunction with symbols in any form of communication.

NOTE

Contrary to this rule, which can be confusing to some extent, it has become commonplace in some branches of industry to use certain word abbreviations due to the repetition of those terms within technical documents. In particular, such terms as kPa(g) are so self-explanatory within a document that their use is tolerated.

---

* The spelling of some SI wording differs between Canada and the United States. The Canadian Standards Association uses the European spelling for the words *metre* and *litre*.

Rule 7: There must always be a space between the last digit and the first letter of the symbol.

Correct Method:

67 kN   59 mA   89 kHz   39.9 kPa(g)

Exception:

No space is left between the numerical value and the symbol when expressing values in degrees.

Correct Method:

Temperature in degrees Celsius: 20°C

Angular measurement: 69° 21′ 24″

Rule 8: Symbols are the same in all languages.

Rule 9:
NOTE

The lowercase l and **L** are both international symbols for liter. To avoid confusion where the l and the numeral **1** could be used in the same expression or document, the United States and Canada have adopted the uppercase **L** as the most appropriate symbol for the liter.

Correct Method:

kg/L   45 L   150 mL (milliliter)

Rule 10: When expressing fractions of an SI unit, decimal designations shall be used, not common fractions.

Correct Method:

1.5   0.255   48.6   347.755

Rule 11: In text and tables, if the numerical value is less than one, a zero shall precede the decimal point.

Correct Method:

0.563   0.987   0.0001   0.889 560 34

Rule 12:   A period shall be used for decimal points.

Correct Method:

See rules 10 and 11 for examples.

Rule 13:   Spaces shall be used instead of commas to separate digits into easily readable blocks of three. Four-digit numbers can be used without a space, for example, 1995 and 2001. However, when columns of figures with four or more digits are tabulated, the spacing should be observed throughout.

NOTE

Caution should be used when interpreting data from countries where the comma is used as a decimal marker.

Correct Method:

2 458.635 89	246.80
15.899 32	1332.00

Rule 14:   A multiplication dot shall not be used as a multiplier in conjunction with numerals.

Correct Method:

36×89 312×24.55 2.5×3.78

Rule 15:   The product of two or more units in symbol form is indicated by a multiplier dot, placed above the baseline. If, however, the printing medium can only provide a period, this is an acceptable alternative.

Correct Method:

N·m   kg·m/s   J/(kg·°C)   kV·A

Rule 16: A solidus or oblique stroke (/), a horizontal line (–), or a negative exponent may be used to express division in a derived unit.

Correct Method:

m/s $\qquad$ $m \cdot s^{-1}$ $\qquad$ $\frac{m}{s}$

Rule 17: The solidus must not be repeated in a single expression. Where required, negative powers or parentheses should be used.

Correct Method:

m/s or $m \cdot s^{-2}$	**NOT**	m/s/s
$m \cdot kg/(s^3 \cdot A)$	**NOT**	$m \cdot kg/s^3/A$

Rule 18: An exponent affixed to a symbol containing a prefix indicates the multiple or submultiple of the unit raised to the power expressed by the exponent.

Correct Method:

$$cm^3 = (cm)^3 = (10^{-2}\ m)^3 = 10^{-6}\ m^3$$

Rule 19: Compound prefixes shall not be used.

Correct Method:

1 nm (nanometer)	**NOT**	1 mμm (millimicrometer)
1 mg (milligram)	**NOT**	1 μkg (microkilogram)

Rule 20: Only one prefix is preferred when forming a derived unit, usually applied in the numerator. Exceptions will depend on practical technical requirements. Note that the kilogram is a *base unit* and is considered as such when used in the denominator.

Rule 21:   The choice of the appropriate multiple of an SI unit is governed by the application. The multiple should be chosen so that the numerical values will be between 0.1 and 1000 when practical.

Correct Method:

12 kN	instead of	12 000 N
3.94 mm	instead of	0.003 94 m
14.94 kPa	instead of	14 940 Pa

Rule 22:   To indicate the division of units in the written form, the word *per* must be used. Do not use the oblique stroke (solidus).

Correct Method:

kilogram per square meter

Rule 23:   When the name of a unit is spelled out, multiplication is indicated by a space.

Correct Method:

pascal second   newton meter

Rule 24:   When writing unit names with prefixes, the prefix name is attached directly to the unit name.

Correct Method:

millimeter   kilogram   decibel

Rule 25:   Unit names and symbols shall not be mixed.

Correct Method:

N·m or newton meter   **NOT**   N·meter

Rule 26:   Only one unit shall be used to designate quantity.

Correct Method:

2.357 kg or 2357 g   **NOT**   2 kg 357 g

SYMBOLS AND UNITS

	Name	Symbol	Quantity

SI Base Units

Name	Symbol	Quantity
meter	**m**	length
kilogram	**kg**	mass
second	**s**	time
ampere	**A**	electric current
kelvin	**K**	thermodynamic temperature
mole	**mol**	amount of a substance
candela	**cd**	luminous intensity

SI Supplementary Units

Name	Symbol	Quantity
radian	**rad**	plane angle
steradian	**sr**	solid angle

SI Derived Units with Special Names

Name	Symbol	Quantity
becquerel	**Bq**	radionuclide activity
coulomb	**C**	electric charge
electron volt	**eV**	unit of energy (nuclear)
farad	**F**	capacitance
gray	**Gy**	energy absorbed from ionizing radiation — any material
henry	**H**	inductance
hertz	**Hz**	frequency
joule	**J**	energy, work, heat quantity
lumen	**lm**	luminous flux
lux	**lx**	illuminance
newton	**N**	force

Name	Symbol	Quantity
ohm	Ω	resistance
pascal	P	pressure, stress
siemens	S	electric conductance
sievert	Sv	dose equivalent
tesla	T	magnetic flux density
volt	V	electromotive force, electric potential, potential difference
watt	W	power
weber	Wb	magnetic flux

SI Prefixes

exa	E	10^{18}	
peta	P	10^{15}	
tera	T	10^{12}	
giga	G	10^{9}	
mega	M	10^{6}	(1 000 000)
kilo	k	10^{3}	(1000)
hecto	h	10^{2}	(100)
deca	da	10^{1}	(10)
deci	d	10^{-1}	(0.1)
centi	c	10^{-2}	(0.01)
milli	m	10^{-3}	(0.001)
micro	μ	10^{-6}	(0.000 001)
nano	n	10^{-9}	
pico	p	10^{-12}	
femto	f	10^{-15}	
atto	a	10^{-18}	

OTHER UNITS WITHIN THE SI SYSTEM

Name	Symbol	Quantity
minute	**min**	time
hour	**h**	time
day	**d**	time
year	**y**	time
liter	**L**	volume
degree Celsius	**°C**	temperature
hectare	**ha**	area
decibel	**dB**	sound
electron volt	**eV**	energy (atomic)
kilowatt hour	**kW·h**	energy
kilometer per hour	**km/h**	speed
metric ton (tonne)	**t**	mass
curie	**Ci**	radionuclide activity
roentgen exposure	**R**	χ or γ radiation
rad	**rad**	energy absorbed from ionizing radiation — any material
rem	**rem**	human radiation dose unit
rem/hour	**rem/h**	human radiation dose rate unit
revolutions per second	**r/s**	rotational speed
revolutions per minute	**r/m**	rotational speed

TABLE 1
Metric Notation Ready-Reference Chart

LENGTH   **Base Unit**

m	mm	cm	dm	km
meter	millimeter	centimeter	decimeter	kilometer

AREA   **Derived Unit**

m^2	mm^2	cm^2	ha
meter squared			hectare

VOLUME CAPACITY   **Derived Unit**

m^3	cm^2	dm^3
meter cubed		

Liquid Volume   **Adopted Unit**

mL	L
milliliter	liter

MASS   **Base Unit**

kg	mg	g	Mg
kilogram	milligram	gram	Megagram

t
metric ton (tonne)

TIME   **Base and Adopted Units**

s	min	h	d
second	minute	hour	day

APPENDIX D

NUMBERS IN TECHNICAL MANUALS

INTRODUCTION

Numbers, both numeral and written, are conspicuous in technical documents because much of the subject matter is quantitative in nature. The use of numerals is generally discouraged in formal nontechnical writing, except where they may be needed for clarity or simplicity. But numerals are favored in technical writing, because they are conspicuous in the text, and thus are easily located for reference, and because they are readily interpreted.

USING NUMBERS IN TEXT

This appendix lists conventions observed in the use of numbers in technical documents.

One-Digit Number

Write out an isolated one-digit number *(one to nine)*.

> A transistor has *three* layers and *two* junctions.
>
> *Bakelite* is only *one* of the trade names of phenol formaldehyde resin.
>
> *Nine* answers are correct; *one* is partially correct.

Do not write out a one-digit number used for calculation or reference.

> Divide the gain factor by 4.
>
> Shall we buy 9 shares at $2 face value?

Numbers Having Two or More Digits

Use a numeral, not words, for an isolated number from 10 upward.

> The servo laboratory is 21 years old.

> The 150th item is missing from the list.

> Use a 50-kg weight.

Hyphenation of Written Numbers

Hyphenate a written-out compound number only when it is in the range *twenty-one* to *ninety-nine*.

twenty-six	his thirty-eighth error
twenty-odd	the fifty-first customer
thirty-nine	our forty-second computation
seventy-five	a thirty-five piece orchestra

Approximations

Write out the number in a statement of approximation.

> about sixteen dollars

> nearly three hundred francs

> a little farther than two miles

Round Numbers

Write out a round number.

a *hundred* pages	a *million* new engineers
a *thousand* cycles	a *hundred and one*
reduced by a *thousandth*	*twenty-odd* miles

Ordinal Numbers

In technical text, write out ordinal numbers from *first* to *ninth,* but use numerals in tables and charts.

our *fourth* contract	the 19th century
the 10th proposal	his *third* attempt
at the 25th closure	their 22nd plant

In a group of ordinal numbers, express each as a numeral if one of them must be a numeral.

> Important odd-numbered harmonics in the square wave are the 3rd, 5th, 7th, 9th, and 11th.

Numbers Separated from a Unit of Measurement

Write out a number under 10 when it is separated from a unit of measurement.

> This fuse will carry *five* and perhaps more *amperes* before melting.

> Remove *three* more *pounds* from the bag.

Numbers Combined with a Unit of Measurement

Hyphenate this combination when it is used as an adjective.

a *five-stage* amplifier	in *10-oz* cups
the *400-Hz* unit	during a *four-day* week
for *50-Hz* operation	on a *5-inch* monitor screen

Numbers Followed by a Unit of Measurement

50 kHz	1 kilometer
1 micron	3 liters
12 feet	4 months and 2 days

Plural of a Number

Form the plural of a written number by adding *s*.

    ones, tens, hundreds, thousands, millions

Form the plural of a numeral by adding *'s*.

    2's   100's   1000's

Number as the First Word in a Sentence

Write out a number when it must be used to begin a sentence; however, avoid beginning a sentence with a number, whenever possible.

Poor:	11 dry runs were made yesterday.
Better:	*Eleven* dry runs were made yesterday.
Ideal:	The crew made 11 dry runs yesterday.

Quantity and Dimension in the Same Expression

When numbers for quantity and dimensions appear in the same expression, write out the quantity and use numerals for the dimension.

    five 144-MHz generators   *twelve* 4-dozen boxes

    sixteen 10-kg weights   the *eighteenth* $500 payment

    two hundred fifty-three shield cans 22 mm in diameter

        two 5-MHz bandwidths

Large Numbers

A large number may be confusing to read (e.g., 167,584,395,010). For clearer expression, numerals and words may be combined.

    $11.3 billion   30 trillion

    580 thousand   0.5 million

    3/4 million   9.3 thousand

Conjunctions in Large Whole Numbers

Do not use *and* between the parts of a large whole number that is written out.

     Incorrect: one hundred *and* fifty

     Correct: one hundred fifty

In the written expression for a sum of money, use *and* to indicate the position of the decimal point (i.e., the division between dollars and cents).

     two hundred thirty *and* 37/100 dollars

     fifty-five dollars *and* 13 cents

Use of Commas in Large Numbers

Do not use a comma in a four-digit number. In a number containing five or more digits, use a comma at the end of each three-digit group counted from the right. Note that this rule does not apply when using numbers within a metric expression (see Rule 13, Appendix C).

1000	10,000	1,000,000
3750	46,385	5,250,130
9000	150,254	111,000,000

Expressing a Sum of Money

Use a numeral for a sum of money.

$1.00	$0.75	$300.50
$250.00	50 cents	$0.10
$30.52	30¢	$1.25

Double Identification in Sum of Money

For security, use a combination of the numeral and written form of the number, but do not use for simple nonmonetary terms.

     Poor: ... are using six (6) generators ...

     Good: The sum of four hundred dollars ($400.00)

Fractions Expressed Numerically

In technical literature, especially in mathematical discussions, fractions are often written numerically.

1½   8¾

3/4   25/1000

12¼   101½

Fractions Written Out

Hyphenate written-out fractions.

one-half   one ten-thousandth

five-eighths   seven and three-quarters

thirty-one   sixty-fourths

Cipher Before a Decimal Point

For clarity, insert a zero before the decimal point in a decimal fraction.

0.0001   0.0535   0.10   0.901

Numbers in Dates

Use plain numbers in dates; the endings *nd, rd, st,* and *th* are not required.

April 2, 1964   August 20

1 December 1968   Nov. 11, 1932

Numbers in Street Addresses

Use numerals for building and street numbers.

37 W. 140th St.   333 Yonge Street

1 College Street   101–103 Exchange Pl.

The street number may be written out, if desired:

86 West Third St.

Caption Numbering

Use numerals for identification in the caption of an illustration, chart, or table.

Figure 12	Fig. 15-1
Plate IX	Fig. III-7
Chart 2 (A)	Table 1

Page Numbering

1	I, i
15	IV, iv
159	980

Serial Numbers

Use numerals in serial numbers.

CRA Report No. 4	Dispatcher 67
Batch 17	Unit 3
Test 10	the year 2001

Percentage Designations

Use a numeral to express a percentage.

15 percent	4%
22½ percent	100%
0.5%	10.8 percent
2.5% less	from 30 to 40%
50 and 75 percent	a 15-percent solution

NOTE

Although it is a preferred practice to write the term percent as one word, it is also common practice for it to be written as two words. In this context, cent applies in the Latin context (centum). The two words actually mean "in every hundred".

Temperature Levels

Use a numeral to express temperature.

20°C	−100°C = 173.16 K
0°F	from −65°C to +100°C
−10 K	−10°C = −8°R

Time Designations

Use a numeral to express time or time intervals.

10 a.m.	9 years*
5:30 p.m.	0.5 microsecond (or 0.5 μs)
half past 8	24 hours*
12 months	45 minutes*

Ratio and Proportions

Use numerals in ratios and proportions.

10:1	4:16 = 5:20
7 to 5	100-to-1 ratio
1:100	10,000 to 1

Designation of Screw Threads

Insert a hyphen between the numeral designating the size of a machine screw and the numerals designating the number of threads.

2-56	6-32
3-48	8-32
5-40	¼-20

---

* Expressions of time such as years, hours, and minutes are not usually fully written out when used in technical documents using S.I. measurements. The form *y*, *h*, and *min*, respectively are preferred. Refer to the tabulation of metric terminology on page 225 if in doubt.

APPENDIX E

ABBREVIATIONS

INTRODUCTION

Abbreviations save time and space. Many abbreviations, such as *a.m., COD, FAA, NBC,* and *CBS,* are in general everyday use; many more are in restricted use in various professions.

This appendix gives abridged lists of abbreviations used in technical writing and some rules for their use.

For approved general language abbreviations, see the current edition of *Webster's New World Dictionary* or the *Concise Oxford Dictionary.*

RULES FOR USING ABBREVIATIONS

1. Do not indiscriminately abbreviate general words. Consult a recent edition of the dictionary when in doubt as to the acceptability of a suggested abbreviation.

2. Use only approved abbreviations. If you must employ a special abbreviation of a private nature, explain its meaning fully the first time you use it in the text.

3. Do not capitalize an abbreviation unless it stands for a capitalized word (*A, Btu, ft L, NEC*). The abbreviations *I.D.* and *O.D.* and those for chemical elements (*Zn, Fe, K,* etc.) are exceptions.

4. Abbreviations usually appear in the singular even when they represent plural words: *kW* (kilowatts) *lb* (pounds), *MHz* (megahertz). Exceptions include *figs.* (figures), *mss* (manuscripts), *pp* (pages).

5. Technical abbreviations use periods only if the abbreviation is spelled exactly like a familiar word from the general language. Thus, *a.m., at., fig., in., no.*

   General abbreviations often contain periods: *Corp., dept., e.g., Ph.D.,* but some do not. The rule here is vague; however, common usage tends to prevail. Write pm instead of p.m., cw instead of c.w. (clockwise or continuous wave). Common abbreviations that have dropped the periods are:

   crt   vhf   uhf   oem   USA   fm

   Note the application of rule 3 in the above examples.

6. Avoid a double-meaning abbreviation if the latter might refer to more than one term in the text. Such abbreviations include *f, fps, min.* Spell out to prevent confusion.

7. Abbreviate the name of a unit of measure only if it follows a number: *150 kg,* but **not** *a little less than a kg.*

8. Abbreviate the word *number* only when it precedes a numeral: *No. 35.*

9. Avoid abbreviations for the names of cities: *Tor, N.Y., L.A., Phila.*

10. Abbreviate a personal title only if it precedes a *full* proper name (first name + last name, or initials + last name): *Dr. F.L. Hudson,* but not *Dr. Hudson.* Otherwise, spell out the title (Doctor Hudson).

11. Abbreviate the name of a month only if the day is given:

    Apr. 2, 1994, but not *Apr. 1994.*

12. Latin abbreviations, such as

    e.g., i.e., viz., ca., Q.E.D., et al., ibid., s.v., vs.,

    should always include the periods.

APPENDIX F

FOOTNOTES

INTRODUCTION

Footnotes are an important part of the complete technical document. Through the use of footnotes, the technical writer, preparing a technical manual,

(a) Provides the name and location of each source of information used by him;

(b) Directs the reader to authoritative sources for further information;

(c) Supplies supplementary information that might, if included in the text, impede the reading; and

(d) Gives credit where it is due.

There is some disagreement among authors and editors as to the quantity and kind of material that should be included in footnotes and how they should be written and punctuated. But there are also many areas of agreement. What is offered here represents, in many instances, a reasonable compromise between the divergent requirements of these areas.

TYPES OF FOOTNOTES

The footnotes used in technical documentation may be classified according to type as *referential* and *informational*.

Referential type footnotes cite a publication, communication, person, or other source from which the writer has obtained information, which she uses as an authority for statements, or to which the reader is directed for further information or for verifying the writer's interpretations and/or conclusions.

[5] Andrew Guthrie, *Vacuum Technology* (New York: John Wiley & Sons, 1963), p.15.

[6] Personal communication from H. J. Mann.

Informational type footnotes introduce material that serves to amplify, explain, or qualify some point but that would unnecessarily impede the reading if it were introduced into the text.

7. Where the approach is similar to that of a previously generated model, it may be reviewed as a stand-alone document.

Note the two styles of reference numbering used; one with the superscript number and the other the full-size number complete with a period.

In general, most technical manual documentation uses the latter style as the footnotes are of the informational type. However, both styles should be used: the first for referential-type material and the second for the informational.

GENERAL PROCEDURES FOR USING FOOTNOTES

1. Use only essential footnotes. Avoid padding a composition with unnecessary footnotes.

2. Do not include important text material in a footnote; it might be overlooked by the reader.

3. Signal the footnote with a superscript (i.e., an exponent or number written above the line) at the end of the sentence, or use a number within parentheses.

These conclusions were reached by the D&D computer group.[1]

or

These conclusions were reached by the D&D computer group.(1)

1. These figures were brought in by A. F. Johnson, who was investigating the claims of Winter and Company.

When several items appear in a single sentence and only one is to be footnoted, the signaling indicator may be placed at the end of the pertinent word or phrase, since, if placed at the end of the line, it might be mistaken to relate to the word after which it appears.

Temperature coefficient is low,[1] dielectric strength high, and dielectric constant low.

4. Place the footnote at the bottom of the page on which it is first signalled. Identify it with the same superscript number used in the text, and if a referential footnote, write this number flush with the left-hand margin of the material on the page.

Set the footnote in a smaller typeface than that of the body of the text. The usual decrease is 1–1½ point sizes.

Separate the footnote from the last line of the text by means of a solid line. If this is the last line of the text on the page, the footnote is set to follows thus:

_____

1. This informational footnote is set using the full-size reference number style.

Start the first line of the footnote immediately after the superscript, but start additional lines flush with the left-hand margin. If the footnote is an informational type, indent the text in the manner of a subparagraph.

5. When several footnotes appear on a single page, arrange them numerically from top to bottom (retain the one style of numbering, however).

[1] Joseph A. Volk, "Gyroscopes," *Data Systems Engineering*, February 1964, p. 28.

[2] This applies to the stainless steel only.

[3] Ralph Young, *Gyroscopes* (New York: Wheelwright Publications, 1964), p. 50.

6. Number footnotes consecutively throughout a document. If the manual is made up of separate sections or chapters, resume the numbering in each division, otherwise the numbers will become unwieldy.

First Referential Footnote

A referential footnote must be presented completely when it first appears. It may then later be referred to in abbreviated form, as explained in the section following. The first referential footnote should contain the following items:

(a) Name of author(s) in the normal order (i.e., first name first).

(b) Title of the work.

(c) Name of publication in which the work appeared (if work is article, chapter, or section).

(d) Facts of publication (city, publisher, date).

(e) Page number(s) of the referenced material.

If the name of the author or editor cannot be determined, substitute the word *Anonymous* or the abbreviation *Anon.* for the name.

The following examples may help clarify the proper form for a footnote.

[1] J. David Fuchs and Stephen W. Garstang, *Electrical Motor Controls & Circuits* (Indianapolis: Howard W. Sams & Co., 1963), p. 113.

[2] Arthur H. Delmege, "Acceleration Switching Servo Valves," *Electromechanical Design*, March 1964, p. 40.

[3] M. A. Holmes and William Neal, "Stresses in Beams," in N. O. Tilton, ed., *Strength of Materials*, 2nd ed. (Chicago: Lakeside Publishing Co., 1964), p. 105.

[4] Anon., "High Temperature Epoxy in Slip Ring Assembly," *Insulation*, April 1964, p. 46.

Note that italics are used for the title of a book (footnote 1) and the name of a periodical or journal (footnote 2), and that the title of an article (footnotes 2 and 4) or chapter (footnote 3) is enclosed in quotation marks. The editor is named separately from the author(s), as in footnote 3.

It is customary to give only the city of publication, but if the city is not well known or several cities have the same name, the state should also be given: Thus *New York, Boston, Chicago* but *Cambridge, England; Summerland, CA; Englewood Cliffs, NJ.*

The issue of some professional journals is identified by volume, rather than year, and it is so listed in footnotes. For this purpose, capital roman numerals designate the volumes, and an arabic number minus the abbreviation *p* designates the page.

[5] David Saxon, "Rubber-Resin Hybrids," *Plastics Society Record*, XXI, 157.

If the list of authors is too long for the space available for the footnote, follow the name of the first author with the Latin abbreviation *et al. (and others).*

[6] A. W. Lo *et al., Transistor Electronics* (Englewood Cliffs, NJ: Prentice-Hall, 1955), p. 10.

Latin Abbreviations

A footnote may be referred to several times. This might involve laborious repetition. To avoid writing out a long footnote several times, standard abbreviations may be used in citing it after the first entry.

The widely used Latin abbreviations for this purpose are *ibid. (ibidem,* "in the same place"), *op. cit. (opere citato,* "in the work cited"), and *loc. cit. (loco citato,* "in the place cited").

Using Ibid.

Use this abbreviation to refer only to the work cited in the immediately preceding footnote. Follow with the new page number concerned.

[7] Glade W. Wilcox, *Basic Electronics* (New York: Holt, Rinehart and Winston, 1960), p. 31.

[8] *Ibid.,* p. 193.

There must be no intervening footnotes, since *ibid.* automatically repeats the name of the work immediately preceding. Also, *ibid.* must be used on the same page with the previous footnote, or on the following page when the two pages face each other with the book open. Otherwise, *op. cit.* must be used.

Using Op. Cit.

When there are intervening footnotes or when the page must be turned between the original footnote and the repetition, use *op. cit.* Precede this abbreviation with the author's last name, and follow with the applicable page number.

[9] Charles E. Van Hagan, *Report Writers' Handbook* (Englewood Cliffs, NJ: Prentice-Hall, 1961), p. 56.

[10] David B. Comer III and Ralph R. Spillman, *Modern Technical and Industrial Reports* (New York: G. P. Putnam's Sons, 1962), p. 123.

[11] Van Hagan, *op. cit.,* p. 179.

Here, footnote 11 is able to skip over the intervening footnote 10 back to footnote 9.

Using Loc. Cit.

Use this abbreviation to refer to the entire footnote to be repeated (i.e., not only the author and work, but the page or passage as well). When *loc. cit.* immediately follows the footnote to be repeated, the author's name need not be repeated.

[12] George Kravitz, *Basic TV Course* (New York: Gernsback Library, 1962), p. 53.

[13] *Loc. cit.*

Some writers prefer simplified English abbreviations to the Latin ones. For example:

[14] Kravitz, p. 798.

[15] *Basic TV Course,* p. 78.

But whether Latin or English abbreviations are used in repeated footnotes, the abbreviated reference should not be repeated to such an extent that the reader must turn back many pages to find the original footnote. A good rule is to repeat the footnote fully after five abbreviated references that run over a number of pages.

Asterisked Footnotes

When only an occasional footnote is used and it is the only one on a page, it is often marked with an asterisk.

The construction requirements for associated power transformers shall conform to the requirements of specification ALS982X53.*

---

* Formerly issued under specification number SPX982X530.

Footnotes for Numerical Matter

Footnotes sometimes must be supplied to numbers in a table or to numerical matter in the text. The use of superscripts for this purpose can introduce confusion; for example, 151^2 might be misread as 151 squared. To reduce the chance of misinterpretation, symbols should be used instead of numbers.

Typical symbols that are not easily confused with numerical data are shown. They are generally used in the order in which they are listed here.

*	Asterisk
†	Dagger
‡	Double dagger
§	Section

After the list has been exhausted, the symbols may next be doubled (e.g., **) and then tripled (e.g., †††). The symbol # is not recommended, since it conventionally stands for *number, pounds,* or *space out.*

APPENDIX G

PUNCTUATION

INTRODUCTION

Formal technical writing requires exact punctuation as a safeguard against ambiguity. Brief rules for the use of each of the punctuation marks predominant in technical writing are given in this appendix. Examples illustrate each use. Those seldom used in scientific and technical prose have been omitted. Two symbols—the *asterisk* and *solidus*—are not strictly punctuation marks; however, they are explained here because they are used increasingly for the purposes illustrated and therefore deserve to be included with the standard marks.

PUNCTUATION MARKS

The Apostrophe (')

Use the apostrophe in the following applications:

1. To indicate possession (noun).

Henry's workbench	Jones' (or Jones's book)
ten companies' answers	the board's decision
workmen's compensation	your technician's fault

Ed's and Jack's notes (individual ownership)
Ed and Jack's notes (joint ownership)

2. To indicate possession (pronoun).

anyone's	neither's
either's	no one's
everybody's	someone's

The possessive of the pronoun *it* is *its* (no apostrophe). *It's* means "it is."

No apostrophe is used in the possessive of a personal pronoun (*his, hers, its, ours, theirs, whose, yours*).

3. To show omission of a letter(s).

    can't   they're
    don't   that's
    isn't   won't

4. To show omission of the first figures in a date.

    the class of '64
    around '45 or '46
    July '51 is a closer guess

5. In the plural of cited words or terms.

    Count the *and's*.
    six *therefore's* in one short paragraph
    Shall I remove the *and/or's?*

6. In the plural of letters.

    a's, b's, and *c's*   silent *g's, h's,* and *p's*
    different *Q's*   sum of the *k's*
    four *x's* and two *y's*   ten *R's*

7. In the plural of numerals.

    1's, 2's, and *3's*   *45's* are all right.
    by *5's* and *10's*   in the West *80's*
    during the *1800's*   two *50's*

The Asterisk (*)

Use the asterisk in the following applications:

1. To signal a footnote citation.

    Response is nonlinear up to 30 V.* The slope may be tailored by proportioning the resultants.**
    (Always locate it *after* the word and a period, if used.)

2. To show omission of letters in a quoted word, especially one that might be deemed offensive.

   b*tch   d*mn   h*ll

Brackets [ ]

Use brackets in the following applications:

1. To identify explanatory material inserted into the text by a second party.

   > The graphs [taken from the screen of a spectrum analyzer] were accepted as exhibits 5 and 9.

2. To indicate an inserted correction.

   > "In 1976, there were a quarter of a million university teachers [actually 232,500] in the United Kingdom."

3. To indicate editorial interpolation of a conjectural sort.

   > The walls of the anechoic room pro[vide] a 20-dB noise reduction.

   > (The original text was mutilated, so the editor supplied "vide"— which is a guess.)

4. To enclose the notification *sic* to indicate a mistake that is not the writer's.

   > The engineer witness reported, "The laboratory for hydrodynamics were [sic] built primarily to test missiles."

   > (The word *sic* indicates that the witness, not the writer, used *were* in the manner shown.)

5. To enclose phrases already containing parentheses.

   > In a special treatment, the sample was exposed to infrared rays [both in the visible (1.3 nm) and invisible (4.3 nm) wavelengths] for 7 hours.

6. To group terms in a mathematical equation containing terms already grouped inside parentheses.

$$E = I\,[R + j\,(\omega L - 1/\omega C)]$$

The Colon (:)

Use the colon in the following applications:

1. To introduce a formal series.

> Three types of resistors were certified as acceptable: composition-carbon, metal-film, and tin oxide.

2. To introduce a phrase that clarifies or amplifies a preceding sentence or clause.

> The plastic compound had simple composition: wood-flour-filled phenolic.

3. To introduce a clause that clarifies or amplifies a preceding clause.

> Our water supply is unpredictable: At times, our combined washing and sluicing involve flow rates as high as 500 gal/min.

4. To separate a word or phrase from related material following it.

> Danger: High Voltage and Toxic Fumes

> Binary Arithmetic: A method of computation involving two-digit numbers.

5. To separate a long appositive from the introductory element of a sentence.

> Many factors combine to reduce lamp life: over-voltage, shock and vibration, prolonged service intervals, and poor design.

6. To stress a short appositive.

> All the samples of fiber-optic cable had the same thickness: 10 mils.

7. Place the colon outside quotation marks.

> The third component is the "rate gyro": This device measures the input rate by comparing the output torque with the torque of a calibrated spring.

8. Between the city of publication and name of publisher in a bibliographical entry.

> Shaw, Henry. *Handbook of English*. Toronto: McGraw-Hill Ryerson Limited, 1979.

9. After a formal salutation.

> Sir: Dear Madam:
> Dear Sir: Dear Miss Lindsay:
> My Dear Sir: Your Honor:
> Dear Mr. Wilson: To Whom It May Concern:

10. Between the numerals expressing time in hours and minutes:

> 2:10 p.m.   10:45   1:30

The Comma (,)

Use the comma in the following applications:

1. To separate elements of a series.

> ... all acids, bases, and salts, may be..
> The abacus, slide rule, or desk calculator are all...

> Capacitors and resistors, relays and switches, and tubes and transistors are found in...

2. To show omission of a word that is understood.

> Highly purified germanium has a resistivity of 60 ohm-centimeters; silicon, 60,000 ohm-centimeters.

3. To provide contrast or emphasis.

   Smith did well in physics, and in engineering.

4. After a dependent clause that introduces an independent clause.

   When the input is asymmetrical, the output is positive-going.

5. Between long independent clauses joined by a coordinating conjunction (*and, but, or,* etc.).

   Employ a superior grade of insulating sleeving wherever space permits its use, and support the wire bundles with clamps spaced not more than 2 inches apart.

6. Between any coordinate clauses (long or short) joined by the conjunction *for*, to distinguish from preposition *for*.

   Germanium is acceptable, for its resistivity is somewhat lower than that of silicon.

7. After a conjunctive adverb (*consequently, however, therefore,* etc.) that introduces a coordinate clause after another clause.

8. To separate a long subject from the verb.

   Silicon is a high-resistivity material; moreover, it is only moderately temperature sensitive.

9. To separate a nonrestrictive word, phrase, or clause from the rest of a sentence (use before and after the nonrestrictive element).

   Word: We are certain that you, too, will be interested in the report.

   Phrase: The spectrum analyzer, a type of frequency-selective VTVM, will identify those harmonics.

   Clause: Forty-one fittings, all of which were pressure-tested this morning, passed final inspection.

10. To separate a nonrestrictive appositive from the rest of a sentence.

    This material is molybdenum disulfide, an additive.

11. To separate a parenthetical expression from the rest of a sentence.

> Quartz, obtained from South America on prime contract, is the selective element of the A-10 filter.

12. To separate a sentence modifier (introductory phrase or clause) from the rest of the sentence to prevent misreading.

> Phrase: Slowly heating, the thermocouple reaches 1300°C.

> Clause: While the silver melted, the workers cleaned a second crucible.

13. After a conjunctive adverb that modifies an entire sentence.

> Therefore, dirt accumulation due to silting between the valve and nozzle causes a gradual flow change.

14. To separate a verbal modifier from the sentence it introduces.

> Discharged, the capacitor is harmless.

15. Before and after a conjunctive adverb that modifies a sentence but does not introduce it.

> These materials are very toxic, however, and must be handled with great care.

16. To separate a direct quotation from the structure that introduces it.

> W. L. Faith reminds, "The status of rules and regulations concerning air pollution currently is in one grand state of confusion."

17. To separate two parallel adjectives not connected by a conjunction.

> lightweight, thermosetting plastics

> A high-speed, broadband preamplifier

18. To separate two parallel adverbs not connected by a conjunction.

> The contacts close quickly, positively.

> Current will rise simultaneously, nonlinearly.

19. To separate contrasted elements.

    We need technicians, not engineers.

20. To separate identical words that occur consecutively.

    What Springfield makes, makes Springfield.

21. To separate the numerals indicating day and year in a date.

    May 10, 1964

22. To separate names of city, county, state, country.

    Albany, NY   New York, USA
    Cleveland, Ohio   Montreal, P.Q., Canada
    London, England   Chicago, Cook County, IL

23. To separate a title from a name that it follows.

    F. E. Carlson, Editor   S. T. Little, M.D.
    Darrell Conway, Ph.D.   Rex Schultz, P.Eng.
    Norman Gardner, Jr.   J. A. Robinson, Pres.
    Lucille Hobson, C.P.A.   Harold Wilson, Architect
    Rolland W. Hawkins, QC   Joseph Hamm, Designer

24. To separate figure groups in a number having more than four figures
    (except in the case of metric notation — see Rule 13, page 220, for
    explanation).

    10,000   230,980   50,000,000

25. After a name in direct address.

    "Colonel, here are the three quarterly reports you requested."

26. After an informal salutation in a letter.

    Dear Bob,   Richard,   My dear Jim,

27. At the end of a complimentary close in a letter.

    Respectfully,   Very truly yours,
    Sincerely,   Yours truly,

28. After *No* or *Yes* when the word introduces a statement.

> "No, that line size will not carry the load."

> "Yes, degassing without baking is now possible."

29. Place a comma inside quotation marks.

> Stray capacitance is a kind of "gremlin," but its effect can be canceled.

The Dash (–)

There are two types of dashes used in publications, the long em-dash (—) and the shorter en-dash (–), with each having specific applications. They are not to be used as or confused with the much shorter hyphen (-), which is explained further on in this appendix.

Use the em-dash (—) in the following applications:

1. To separate an abrupt interrupting statement.

   > Solid-state research—at least in the 1970–1990 decades—surpassed all other activities.

2. To separate independent clauses abruptly.

   > Dip the seed crystal into the melt—pull it out slowly.

3. To separate a non-restrictive phrase that already contains commas, from the rest of a sentence.

   > Sensors of several kinds—accelerometers, strain gauges, vibration pickups, and pressure pickups—can be mounted on the rotor of the test helicopter.

4. To separate a compound appositive from the rest of a sentence.

   > All qualified professionals—engineers, architects, and surveyors—are licensed by the appropriate provincial bodies.

5. To separate a phrase or clause that summarizes a series given in a preceding clause.

   > Phrase: Muffle, tube, and crucible laboratory furnaces are available for combustion study—all of them.

Clause: Fractional horsepower motors, speed controls, and gear boxes—all are built at our older Cambridge plant.

6. To separate a credit citation at the end of a quotation.

All types of electronic units can be broken down into three sections: Sensory Devices, Circuits (including components), and Load Devices—W. P. Wilcox, *Basic Control System Electronics*, p. 134.

Use the en-dash (–) in the following circumstances:

1. As a substitute for the word "to" in continuing numbers.

1990–1995	10.30 p.m.–1.30 a.m.
pp. 360–400	Figures 10-1–11-15
30°F–63.4°F	6.24–32 and Table 6-4

2. As the equivalent of "and" or "to" in two-word expressions.

space–time   series–parallel circuit
metal–glass bond   nickel–cadmium cells

3. To join the names within organization titles.

Baldwin–Lima–Hamilton and Associates
Bendix–Balzers Vacuum, Inc.
Granville–Phillips Co.
Bentley–Harris Manufacturing

The Ellipsis (...)

Use the ellipsis in the following applications:

1. To show omission of material from a quotation.

According to Mayor Starkie, "...complete control of main artery traffic flow is imminent."

The neon diode...is a simple, inexpensive switching element.

Dr. Tucker says, "Development of adequate theories for explaining bulk action in organic semiconductors depends on rigorous mathematical analysis...".

2. To show continuation of a series of elements beyond the last one mentioned.

> The use of conjunctive adverbs (*accordingly, however, moreover, nevertheless,...*) is more formal than simple.

3. To indicate a number of terms not shown in a mathematical expression, such as a progression or series, after the last term given.

$$1 + 2 + 3 + 4 + \cdots + n \qquad\qquad 1, 2, 3, \ldots$$

The Exclamation Mark (!)

Use an exclamation mark in the following applications:

1. At the end of an exclamatory sentence.

> Both missiles exploded!

> Ready, aim, (pause), fire!

2. After a strong interjection.

> Good grief!   So what!   Well done!

3. Place an exclamation mark outside quotation marks when the quotation itself is not an exclamation.

> The investigating officers were surprised to hear him admit, "I have had no contact with him"!

4. Place an exclamation mark inside quotation marks when the quotation itself is an exclamation.

> "It's been damaged beyond repair!"

5. To indicate factorials in a mathematical expression.

$$2! \qquad n! \qquad 1/n!$$

NOTE
The exclamation mark should be used sparingly in any type of writing; over-use only lessens the effect.

The Hyphen (-)

Use a hyphen in the following applications:

1. To join parts of some compound nouns.

   dyne-centimeter   watt-hour
   foot-candle   T-section

2. To join parts of a compound adjective that precedes the word it modifies.

   capacitor-start motor   *short-term* stability
   heavy-duty suspension   *solid-state* devices
   leak-tight coupling   *45-inch* TV screen
   right-angle drive   *thin-film* semiconductors

3. To join parts of an adjective phrase.

   air-to-air missile
   easy-to-read instructions
   state-of-the-art hardware
   10-to-15-page manuals

4. To join parts of some compound verbs.

   air-condition   sun-dry
   surface-grind   water-soak
   self-ignite   wire-record

5. To separate the prefix *all-, ex-,* or *self-;* or suffix *-el,* from root word.

   all-around versatility   *ex*-president
   self-generating photocell   Mayor-*elect*

6. To separate a prefix from a proper name.

   anti-Quebec   pre-Michelson
   ex-Mayor Drapeau   trans-Canada

7. To join names or initials in some names of organizations.

   Q-Max Corporation   W-K-M Division

8. To prevent confusion in reading.

> *heavy-hydrogen atom* versus heavy hydrogen atom
> *man-eating fish* versus man eating fish
> *two-ton trucks* versus two ton trucks

9. To separate syllables to avoid confusion with identically spelled word.

> *re-collect* versus recollect
> *re-cover* versus recover
> *de-crease* versus decrease
> *de-face* versus deface
> *pre-serve* versus preserve

10. To prevent awkward combination of letters in compound word.

contra-acting	re-encapsulate
electro-optical	shell-like
pro-oceanic	semi-inductive

11. To prevent ambiguity when multiple modifiers are used.

English-speaking people	forty-year-old scotch
power-take-off gear	store-to-door delivery

12. To form written compound numbers from 21 through 99.

twenty-one	seventy-eight
forty-five	eighty-seven
fifty-six	ninety-one

13. To form compound words where the modifier changes the last term.

cross-section	self-sustaining
cross-feed	half-exposed

14. To split word at end of line.

> Always break at the end of a syl-
> lable as is done in this sentence.

(Split a word only at the end of a syllable. Do not split a one-syllable word. Split a hyphenated word at the hyphen, if possible. Do not split a word so that a single letter is carried over to the next line.)

15. Since so much indecision arises regarding hyphenation in technical writing and editing, the following rules seen helpful to add to those already given.

    (a) Do not ordinarily hyphenate a compound noun:

      A photocell is a long-lived source of *direct current* [not *direct-current*].

      A few such nouns are regularly hyphenated. These include the combined names of two dissimilar devices operated together or performing two functions (*digital-voltmeter, motor-generator, etc.*) and units of measurement composed of the names of two dissimilar units (*kilowatt-hour, ampere-turns, foot-candles,* etc.).

    (b) Hyphenate a compound modifier only when it *precedes* the word it modifies, not when it serves as the predicate adjective:

      This is a *high-frequency* power supply
                      but
      The power supply is *high frequency*.

    (c) Hyphenate a compound modifier consisting of *well* + verb only when the modifier *precedes* the word it modifies. Do not hyphenate when it follows the modified word:

      Correct: This material is a *well-cured* resin.

      Correct: The Home Hardware's Markham warehouse store is *well stocked*.

    (d) Do not insert a hyphen between an adverb ending in *-ly* and the word it modifies.

      Incorrect: All that was found was the *badly-burned* tail section of the plane.

      Corrected: All that was found was the *badly burned* tail section of the plane.

    (e) When in doubt about hyphenation, look up the word in the latest edition of a dictionary. Also consult the glossary (lexicon) of your profession. When the dictionary and the glossary disagree, obey the glossary.

Parentheses ( )

Use parentheses (in pairs) in the following applications:

1. To enclose a nonrestrictive element inserted in a sentence.

    Ethylene-vinyl acetate copolymers (introduced by U.S. Industrial Chemical Co.) are competitive with vinyl and rubber.

2. To enclose a clarifying or amplifying statement within a sentence.

    Members of several national societies (ASME, IEEE, NSPE, and the SAE) presented papers at the meeting.

3. To enclose a cross-reference introduced in a sentence.

    Machrone discusses IBM's commitment to OS/2 (see *PC Magazine*, Volume 11, June 1992, p. 87) in his editorial.

4. To enclose added information.

    Rensselaer Polytechnic Institute is in Troy (NY).

5. To enclose a question mark indicating doubt or uncertainty.

    Thales, born 640 (?) B.C., reported on the electrification of amber.

    One engineer (?) will handle all their research.

6. To enclose a sum repeated, for accuracy, in figures.

    Will the maximum appropriation be six thousand dollars ($6000)?

7. To enclose numerals or letters designating the elements of a list.

    The ceiling cell contains (1) power conduit, (2) compressed air lines, (3) intercom cable, and (4) air ducts.

    Stability may be improved by (a) aging the neon lamps, (b) voltage-matching the aged units, and (c) operating them at a low-duty cycle.

8. To enclose a number to indicate that it is multiplied by an adjacent number.

    $$234 = 2(10^2) + 3(10^1) + 4(10^0)$$

9. To group terms in a mathematical expression.

$$y + dy = (x + dx)^2$$

10. Punctuation and capitalization of parenthetic element.

   (a) When a parenthetic element is inserted in a sentence, do not use an initial capital or final period.

   The four-layer diode (see Chapter 4) is a PNPN device.

   (b) When the parenthetic element is outside the sentence, use an initial capital and final period.

   Only passive stabilizing elements were used. (Thermistors were the most satisfactory.)

11. When quoting references.

   Figures 4(a) and 4(b), *not* Figures 4(a,b).

The Period (.)

Use the period in the following applications:

1. To mark the end of a declarative or imperative sentence.

   Declarative: An AND circuit delivers outputs only when a pulse appears simultaneously at each of its input terminals.

   Imperative: Degrease the aluminum wire before you use it.

2. After certain nontechnical abbreviations.

anon.	Dr.
B.S.	Ph.D.

3. After technical abbreviations that might otherwise be read as familiar words.

bar.	in.
cap.	at.
fig.	no.

4. As the radix point in a number.

Binary point:	1110110.1011
Octal point:	3742.1526
Decimal point:	10.7503

5. To mark the division of dollars and cents in a written sum of money.

$1.00 $5.98 $1150.61

6. Place a period inside quotation marks.

Eliot reports, "Low-frequency radio has a long record of poor performance at this location."

The Question Mark (?)

Use the question mark in the following applications:

1. To indicate the end of an interrogative sentence.

What is the efficiency of the gas turbine?

2. To suggest doubt or uncertainty.

First commitments were made in July (?).

Archimedes (287?–212 B.C.) was an early observer of the effects of specific gravity.

3. Place a question mark inside quotation marks when a question is quoted.

Pridham asked, "What is the level of scientific creativity in your division of the company?"

4. Place a question mark outside quotation marks when the quotation is part of the question.

Did the inspector not say, "The care of classified documents at this plant is abominable"?

5. Use only one question mark (inside quotation marks) with a double question (a question that quotes a question).

Did the officer ask you, "Will you sign the statement?"

Quotation Marks (" ")

The preferred style for quoted material is (" and "); however, for mere emphasis of words and phrases, use the " style.

Use quotation marks in the following applications:

1. To identify a direct quotation.

    > The editor said, "Send 200-word abstracts of all papers not later than June 15th."

2. To identify an unusual or nonstandard term.

    > A single-stage "filtrap" provides 100 Hz bandwidth at 100 kHz center frequency.

    > They took the car out and "blew its nose" on the freeway, removing most of the carbon.

3. To identify a definition.

    > As used in this report, *pickup* means "a sensor of any kind."

4. To identify the title of an article, chapter, or section quoted from a book, report, or periodical.

    > Robert A. Leonard, "Design for a Data Exchange System," *Electronic Industries*, May 1964, p. 86.

5. Use single quotation marks with a word or phrase included within a quotation.

    > The contract specifically states, "No patent liability is assumed for 'gadgets' using the circuits shown herein."

6. Place a colon outside quotation marks.

    > Several things may be meant here by "light": radiant energy, visible electromagnetic radiation, and control beams.

7. Place a comma inside quotation marks.

    > The "spark pump," a piezoelectric ignition device, is described in the paper by J. P. Arndt.

8. Place a period inside quotation marks.

    > The label warns, "Do not inhale the vapors."

9. Place a question mark inside quotation marks when a question is quoted.

> In technical reports, avoid rhetorical questions such as "What is the implication of these findings?"

10. Place a question mark outside quotation marks when the quotation is not part of the question.

> Why didn't he say "The witness actually broke down"?

11. Place a semicolon outside quotation marks.

> See "Trends in Air Pollution Control"; this article by W. L. Faith takes a good look at the problems of smog control.

The Semicolon (;)

Use the semicolon in the following applications:

1. To separate the elements of a series when the elements already contain commas.

> ...diodes, transistors, and rectifiers; thermistors, varistors, and varactors; and photocells, magnetoresistors, and lumistors.

2. To separate coordinate clauses not joined by a conjunction.

> The pressure increased to the danger point; automatic control started immediately.

3. To separate the main clause of a sentence from a long or loose clause introduced by a coordinating conjunction.

> These magnets are plastic encased; and it has been found that the cost of the process for manufacturing in large numbers can be substantially reduced through use of injection molding.

4. To separate clauses joined by a conjunctive adverb.

> It is fairly easy to make a glass-to-metal seal; however, molten glass will not fill small crevices.

5. In compound sentences containing commas.

> A photoelectric relay provides simple circuitry; but, if you desire a reasonably tamper-proof system, use radio control.

6. Before an introductory word, phrase, or abbreviation (*e.g., i.e., for example,* etc.)

> Some number systems surpass the decimal system in adaptability to electrical counting circuits; for example, the binary system is easily handled by on–off switching.

7. Place a semicolon outside quotation marks.

> John F. Brinster surveys the art in "Multi-channel Telemetry Communication Devices"; here he devotes some space to electro-mechanical vs. electronic communication.

Oblique Stroke or Solidus (/)

Use the oblique stroke in the following applications:

1. To stand for the word *and* in some expressions.

ac/dc circuit	*message/data* switching
adhesive/sealant	research/development
breakdown/failure	*transmit/receive* switch

2. To stand for the word *or* in some expressions.

and/or	*open/short* tester
gain/loss figure	*read/write* amplifier
go/no-go basis	*reflection/refraction* effects

3. To stand for the word *per* in some expressions.

ft/s/s	0.1%/°C
lb/min	10 μV/day drift

4. To stand for the word *to* in some expressions.

L/C ratio	*binary/decimal* encoder
dc/ac inversion	*signal/noise* ratio

5. To show where a new page starts in the original version of material being copied.

> "In general, lengths measured with the microscope should be suspect if /84/ they are less than about 10 times the resolution limit of the objective...."

6. As the fraction bar in "on-line" fractions.

1/2	15/16	*Y*/5
3/4	*f*/2	$(a + b)/(c + d)$

GLOSSARY

This glossary includes material from this book plus other publishing terms which the reader may come across in the process of preparing a technical manual.

ACKNOWLEDGEMENT: The part of a book that gives credit or thanks.

ADDENDUM: An addition to a manual, as distinct from an appendix.

ALPHANUMERIC: Consisting of letters, numbers, and symbols (generally, those available on a typewriter).

AMMONIA-VAPOR (MACHINE): Used to develop exposed light-sensitive paper to produce blueline on white prints. The rugged nature of the printed paper is useful for rough-handling situations. Capable of producing very large drawings. Also known as DIAZO.

APPENDIX: Material of interest to the reader which is not part of, and usually follows the text.

APPLE MACINTOSH: A commercial brand of desk-size computer used mainly for desktop publishing and/or the preparation of illustrations.

ARABIC NUMERALS: The familiar digits used in arithmetical computation.

BACK MATTER: Material following the text, includes indexes, bibliographies, glossaries and appendixes. The pagination style should follow the body of the text.

BIBLIOGRAPHY: A list of publications of interest to the reader.

BITE: Acronym for Built-In Test Equipment.

BLOCK DIAGRAM: A simplified diagram illustrating the principles of operation of an equipment.

BOLDFACE: Darker than normal typeface. See also, TYPESTYLES.

BOND PAPER: Paper made chiefly for use in offices and as writing paper.

BOOK PAPER: Paper made principally for the manufacture of books, pamphlets, and magazines as distinguished from newsprint, writing and cover stock.

BRACKETS: A device for enclosing material [thus].

BROCHURE: A pamphlet bound in the form of a booklet.

BULK: The thickness of paper in number of pages per inch; also used loosely to indicate the thickness of a book, excluding the cover boards.

CAD/CAM Computer-Aided Design/Computer Assisted Manufacturing.

CAMERA-READY COPY: Artwork, type proofs, typewritten material, etc., ready to be photographed for reproduction without further alteration.

CAPS: An abbreviation for CAPITAL LETTERS.

CAPS AND SMALL CAPS: Two sizes of capital letters which are available in the same font and often used together. FULL CAPITALS appear like this and the SMALL CAPITALS like this.

CAPTIONS: The titles or descriptions of illustrations and tables.

CASE: A cover or binding, made by machine or by hand and usually printed, stamped, or labelled before it is glued to a book.

CASE BINDING: In case binding, sewn signatures plus endpapers are enclosed in a rigid cover. Thread, passed through the fold of each signature locks it at the back. Side sewing passes the thread through the entire book from the side. Side wiring is essentially the same, except that wire staples are used instead of thread. At this stage, the book is referred to as unbound signatures. A hinge of heavy gauze is glued to the spine, and the signatures are attached to the rigid case by the endpapers and the flaps of the hinge.

CATHODE-RAY TUBE: A vacuum tube with a screen at one end, illuminated by means of an electron beam controlled by magnetic devices at the other end. A television picture tube and the video screen of a computer monitor are common examples. Abbreviated CRT.

CENTERED DOT: A heavy dot (•) used as an ornament before a paragraph. A lighter centered dot (·) is used in mathematical composition as a multiplication sign. Also known as a BULLET.

CENTRAL PROCESSING UNIT: A computer, or section of a computer, that performs all the main operating functions of the system and generally contains the main data-storage area. Abbreviated as CPU.

CHAPTER HEADS: The numbers of the chapters (not the titles).

CHARACTER: A letter, numeral, symbol, or mark of punctuation. In printing type, characters vary in width. On an ordinary typewriter characters are all the same width.

CLIP ART: Commercially prepared illustrations, usually gathered in sets, and available for use with many programs used on desktop computers.

COATED PAPER: Paper to which a surface coating of clay or other opaque material has been applied. Papers may be coated one side or two sides. Finishes range from matte (dull-coated) to very shiny (gloss enamel).

COLLATE: In bookmaking, to arrange the folded signatures of a book in proper sequence for binding. In manual preparation, refers to the assembly of sections, chapters pages, foldouts, index dividers and front matter in correct order.

CONDENSED (TYPE): The typeface with characters narrower than normal, permitting more material to be set in a line of the same width.

COPY: Typescript, original artwork, photographs, etc., to be used in producing a printed work. (see also, CAMERA-READY COPY)

COVER: The outside binding or case of a book or manual, whether it be paper, plastic or cloth, fixed or loose.

COVER MATERIAL: Flexible material, such as leather, cloth, paper, or plastic, used to form the cover in CASE BINDING.

COVER STOCK: Paper, generally thicker than book paper, used for the covers of pamphlets, brochures, paperback, etc.

CPU: See CENTRAL PROCESSING UNIT.

CPM: (Critical Path Monitoring). A flow diagram with which the progress of a job can be monitored against the estimated progress.

CROP: To cut down an illustration, such as a photograph, in size to improve the appearance of the image by removing extraneous areas. Cropping is performed not by physical cutting but by masking, and crop marks are placed on the photograph or drawing as a guide to the printer's cameraman.

CRT: See CATHODE-RAY TUBE.

CUT-AND-PASTE: The technique of trimming pre-printed text and illustrations to fit onto a layout sheet in preparation for reproduction. The term originally applied to the manual operation but is now used to denote similar procedures carried out using computer software.

D-RING: A type of binder holding ring that permits the edges of the contents to sit square in the binder.

DESCENDER: The part of such letters as p, q, and y that extends below the base line, or bottom of the capitals.

DESKTOP PUBLISHING: A term used to denote the preparation and publishing of a variety of documents using a personal (desktop-size) computers, laser printers and sophisticated software programs.

DIE, STAMPING: A die of brass or other hard metal used to stamp the case of a book with titles or logotypes.

DIE-CUTTING: The process of cutting regular or irregular shapes out of paper by the use of specially fashioned steel knives. The result may be a hole in the cover of a manual to show a title block on the first page.

DISPLAY TYPE: Type that is larger than the body type used for setting text of a printed work. Display faces are used for title pages, chapter opening, subheading, etc., in a book or journal, for headlines in advertising, and so on.

EDITOR: One who prepares material for composition, including spelling and grammar, checking for consistency of style, and improving legibility when required.

ELITE TYPE: Typewriter type that runs twelve characters to be inch (also called 12 pitch). See also, PICA TYPE.

EN-DASH: A dash, the length of the typeface "n", used to substitute for "from" and "to" in numbers and two-word expressions. Also used as hanging indentation lead-ins.

ENDLEAVES: A blank leaf pasted to the inside of the front and back covers of a book. Also called ENDPAPERS.

EM-DASH: A dash, twice as long as an en-dash, used to space sections of text and thoughts.

EXPLODED VIEW: A drawing of a piece of equipment, with all individual parts shown, and in alignment with their origin. Used for repair and/or part location.

FIGURES: (Also, FIGURE NUMBERS) Numerals used to designate illustrations.

FLOWCHART: A drawing showing the progress of a system's operation, with symbols used to represent operations, data or material flow. Lines and arrows are used to represent relationships.

FLUSH: In typesetting, lines set flush left are aligned vertically along the left-hand margin. Flush right means the opposite. See also RAGGED RIGHT.

FLYLEAF: Any blank leaf at the front or back of a book, except the endleaves pasted to the inside of the cover.

FOLDER: A machine that folds printed sheets into signatures for binding, often attached directly to the press at the delivery end.

FOLDING GUIDES: Small lines printed on the edges of large sheets of paper to indicate where pages are folded (usually found on foldout drawings).

FOLDOUT: An oversize page, often an illustration or a table, folded to fit within the trim size of the book, with the title immediately visible before unfolding.

FOLIO: A page number, often placed at the outside of the running head at the top of the page. If placed at the bottom of the page, the number is a drop folio. A folio counted in numbered pages but not printed (as on a blank page at the end of a section) is a blind folio; any folio printed is an expressed folio.

FONT: A complete assortment of a given size of type, including capitals, small capitals, and lowercase, together with ligatures, punctuation marks, ligatures, and the commonly used signs and accents. Many special signs and accents are available but are not included in the regular font. The italic of a given face is considered a part of the equipment of a font of type but is spoken of as a separate font.

FORMAT: The shape, size, style, and general appearance of a manual as determined by type, page size, margins, and so on.

FRONTISPIECE: An illustration that faces the title page or used as part of the title page.

FRONT MATTER: Material preceding the text, including title page, foreword, preface and contents. In technical manuals, this may also include additional material such as safety information, specification sheet, warranty, and so on. Folios are usually numbered in lowercase roman numerals.

GRAIN: Paper supplied in sheets may be cut grain long (grain running the long way of the sheet) or gain short (grain running the short way). May affect folding characteristics.

GRAPHIC: An individual piece of artwork. A term commonly used for computer-generated illustrations.

GRAPHIC ARTIST: A professional artist who produces artwork either by the traditional pen and ink method or on a computer using sophisticated drawing software.

GUIDELINES: Lines printed or drawn on layout sheets to facilitate the straight placement of a printed work and illustrations.

GUTTER: The two inner margins (back margins) of facing pages of a book.

HALFTONE: A process whereby an image, such as a photograph, is broken up into a pattern of dots of varying size from which a printing plate is made. When printed, the dots of the image, though clearly visible through a magnifying glass, merge to give an illusion of continuous tone to the naked eye. Reproductions of photographs in printed matter, whether letterpress or offset, are called halftones.

HANGING INDENT: The first line of a paragraph is set to the margin and the following lines are indented, whether as straight text or by subparagraphing.

HARDWARE: In computer terminology, machinery, circuitry, and other physical entities, as distinct from SOFTWARE which are operating programs.

HELVETICA: A very common plain unadorned typeface.

HINGE: In a bookbinding, the connection between the covers and the book proper; in hand binding it's strength is due to the tapes or cords to which the signatures are sewn, in a case binding, to a strip of gauze.

INDENT: To set a line of type so that it begins or ends inside the normal margin. In paragraph-style indention the first line is indented from the left-hand margin and the following lines are set full measure. In hanging indention, the first line is set full measure and the following lines are indented.

INFERIOR FIGURE: A small numeral that prints partly below the base line: A_2. See also SUBSCRIPT.

INDEX: A list of significant names and terms used in the text, with the folios of pages on which they appear.

INSTALLATION (MANUAL): The detailed procedure for the preparation and installation of a piece of equipment.

ITALIC: Slanted text. See also, TYPESTYLES.

ISO: (International Standards Organization). Governing body for the metric system of measurements (also called SI).

JUSTIFY: To adjust the word and/or letter spacing in lines of type in order to achieve straight margins at both left and right edges of the text.

KERNING: The fine adjustment of spacing between characters and/or certain letters.

KRAFT: Brown paper, used chiefly for wrapping paper, made from unbleached sulfate pulp.

LATIN ALPHABET: The ancestor of our alphabet, consisting of twenty-one letters (j, u, w, y, and z lacking). It is the parent of alphabets used in printing western European languages, including the Old English, German Fraktur, and Irish forms of letters. Latin is also used to distinguish an alphabet like ours from such forms ad the Greek, Cyrillic, and Semitic alphabets.

LAYOUT: A designer's conception of the finished job, including spacing and type specification.

LEADERS: A row of dots, evenly spaced, designed to carry the reader's eye across the rows of a table, from the chapter title to its page number in a table of contents, and so on.

LEADING: Extra spacing between lines of type, in addition to that provided by the type itself.

LEGEND: The title or description of an illustration or table (see also, CAPTION).

LETTERPRESS PRINTING: Printing from a raised surface in which the ink is transferred directly to the paper.

LEXICON: A dictionary of specialized terminology for a specific engineering or scientific field, for example, nuclear energy.

LIGHTFACE: The ordinary variety of roman or italic type, as distinct from boldface. See also TYPESTYLES.

LINE DRAWING: An illustration containing solid black lines on a light background.

LOGO: (see LOGOTYPE).

LOGOTYPE: More commonly known as the "logo"; one or more words, or other combinations of letters used for company names, trademark, graphic, and so on.

LOOSELEAF: (BINDING) A mechanical form of book binding to provide easy removal of pages.

LOWERCASE: The uncapitalized letters of a FONT (abbreviated lc).

MACHINE COPY: A copy of anything made on an office copying machine, such as a Xerox machine; often used as proof.

MAINFRAME: In computer terminology, a large CENTRAL PROCESSING UNIT, as distinct from input and other devices attached to it. The term is commonly reserved for powerful scientific and business-oriented computers.

MAINTENANCE (MANUAL): A section of a manual that deals exclusively with the servicing and/or repair of an equipment.

MANUSCRIPT: An author's copy of a work as submitted to a publisher or printer.

MARGINS: The white space surrounding the printed area of a page, called a variety of names such as the back, or gutter, margin; the head, or top margin; the fore-edge, or outside margin; and the tail, foot, or bottom margin.

MARKUP: The process of marking manuscript copy for typesetting with directions for use of type fonts and sizes, spacing, indention, and so on.

MEASUREMENT: The printer's basic unit of measurement is the point, approximately 1/72 of an inch; 12 points equal 1 pica, approximately 1/6 of an inch. Within a font of type of one size the printer commonly measures by ems.

MECHANICAL BINDING: A category of binding style for books and pamphlets in which the spine fold is trimmed off and the leaves punched to accept some device that holds them together. Spiral, plastic (or comb), post, and ring bindings are examples. Most common method of binding for technical manuals.

MIL-SPEC: A specification based on US government military requirements.

MYLAR: Trademark name for a Dupont polyester film.

OCR: Optical character recognition.

OEM: Original equipment manufacturer. One who produces subassembly items in quantity.

OPERATION (MANUAL): A section of, or a specific manual detailing how an equipment is to be used or operated.

OFFSET (PRINTING): An indirect method of printing in which the ink is transferred to an intermediate surface and from that to the paper.

OVERHAUL (MANUAL): The procedure to bring an item of equipment up to its original operating status.

PAGE MAKEUP: The manual arrangement of text and illustrations, photographs etc., into page lengths. In computerized typesetting, the electronic assembly of page elements into a completed page, as viewed on a computer monitor.

PARENTHESES: A form of brackets () used to contain amplifying or explanatory elements in sentences. Also for enclosing numerals and letters in lists.

PAGINATION: The numbering of pages in a book.

PASTE-UP: The camera-ready assembly of elements of a page on a card in preparation for reproduction.

PCB: Printed circuit board.

PERFECT BINDING: Holding pages together with an adhesive along the back edge after the folds are trimmed off.

PERT: (Program Evaluation and Review Technique). A management control method of defining what must be done to accomplish a desired objective on time.

PHOTO-INVISIBLE (MARKING): Marking layout lines and instructions on camera-ready material with light blue lines and notations. This color does not reproduce when phototypset.

PHOTO-TYPESETTING: Composition by a photographic process.

PICA: A unit of typographic measurement, equal to 12 points.

PICA TYPE: The typewriter type that runs 10 characters to the inch. See also, ELITE TYPE.

PLASTIC BINDING: A type of mechanical binding; also called comb or Cerlox binding.

PMT: (Photo-Mechanical Transfer). A photographic process whereby a large or small drawing can be reproduced, as a positive, to a specific size on material suitable for use on layout sheets.

POINT: The printer's basic unit of type measurement – 0.0138 inch (approximately 1/72 inch).

PROOFREADER'S MARKS: An almost universally accepted system of marking errors on proofs.

RAGGED RIGHT: Set with the right-hand margin unjustified. See also JUSTIFY.

RAG PAPER: High-quality paper made from cotton rags, processed to form a pulp.

REAM: The number unit on the basis of which paper is handled, usually 500 sheets.

RECTO: The front side of a leaf in a book; a right-hand page. A preface, an index or a chapter normally starts on the recto page. See also, VERSO.

REGISTER MARKS: Crosses or other forms of makings used to align original copy prior to photography. Essential in multi-color layouts.

ROMAN: The ordinary type style, as distinguished from italic.

ROMAN NUMERALS: Numbers formed from traditional combinations of roman letters, either capital (I, II, III, IV, etc.) or lowercase (i, ii, iii, iv, etc.). See also, ARABIC NUMERALS.

RULE: A line used as a separator.

RUNNING HEAD: A heading placed at the top of a page as an aid to the reader. It may contain the number and the title of the chapter.

SADDLE STITCH: A method of stapling small booklets and magazines along the fold.

SAN SERIF: A typeface that is unadorned. See also, TYPESTYLES.

SCALE: To scale an illustration is to calculate (after cropping), the proportions and finish size of the reproduction, and the amount of reduction needed to achieve this size.

SCANNING: An means of transferring text or illustrations from different sources (books, original drawings, and so on) to a computer program.

SCORING: Making an impression on paper to facilitate folding.

SCREEN: A dot pattern in the printed image produced by such a screen.

SERIF: A short light line projecting from the top or bottom of a main stroke of a letter. Originally produced in handwriting, at the beginning and finishing stroke of the pen nib.

SEWING: See CASE BINDING.

SI: Common abbreviation for the metric system of units.

SOLIDUS: (also known as oblique stroke, slant or slash). A type character consisting of slanted line (/).

SIGNATURE: Sheets of a book, printed and folded ready for sewing. It is often 32 pages but may be a small as eight pages, depending on paper thickness. If the paper is very thin, the additional folding ability permits the signature to be 64 pages.

SOFTWARE: Computer programs.

SPINE: The part of a book binding visible when the book is shelved. The title should read normally when the book is lying face up on a flat surface.

SPIRAL BINDING: A type of MECHANICAL BINDING.

STOCK: Paper to be used in the printing of a manual.

STYLE MANUAL: A reference book for authors, editors and publishers providing standards for the editing and publishing process.

SUBHEADS: Section titles or headings subordinate to chapter heads.

SUBSCRIPT: A small symbol that prints partly below the base line of the text; H_2. See also, INFERIOR FIGURE.

SUPERIOR FIGURE: A small numeral that prints partly above the X-HEIGHT; A^2. See also, SUPERSCRIPT.

SYMBOL: A special character used to denote a mathematical term, for example, $\equiv$.

TEMPLATE: In desktop publishing, page designs or layouts created to facilitate the production of a variety of documents without time-consuming design time.

TEXT: The body of a book, not including front or back matter, illustrations, tables and so on.

TIMES ROMAN: The most common typeface in use today. Can be found in newspapers, magazines, etc. This text used in this book is Times Roman.

TYPOGRAPHY: The art of arranging printed type.

TYPESET: To set a manuscript into type; to compose.

TYPESTYLES: The type commonly used in books and all classes or reading matter is known as roman. Although all roman types are essentially the same in form, there are two well-defined styles. The *old* style, characterized by strength with strokes of comparatively uniform thickness and with an absence of weak hairlines. The serifs are rounded, and the contour is clear and legible. Caslon is an example of an old-style face.

The second style is called *modern* and is characterized by heavier shadings, thinner hairlines, and thin, straight serifs. Bodoni is an example. Aside from the roman, there are four general classes, known as italic, script, gothic and text. Boldface versions of all the commonly used faces, in both roman and italic are available. These may also be found in extended and condensed versions.

The slanting letter mainly used for emphasis and display is known as *italic*. It is cut to match all roman typefaces, and a font of roman type for book and magazine work would be considered incomplete without a corresponding font of italics.

Script types are imitations of the old style copperplate handwriting. The widest use is for formal announcements, invitations and stationary.

Gothic or *sans serif* is perfectly plain, with lines of uniform thickness and without serifs. It is sometimes known as block lettering.

TYPESIZE: Generally the distance from the top of the highest character in a font to the bottom of the lowest.

UNJUSTIFIED: Lines of type that do not have an even right-hand edge. Also known as RAGGED RIGHT.

UPPERCASE: Capital letters.

VERSO: The left hand or rear of a page of a book (see also, RECTO).

BIBLIOGRAPHY

1. *AIP Style Manual*, 4th Edn., New York: American Institute of Physics, 1990.

2. Brusaw, C.T., G. J. Alred and W. E. Oliu. *Handbook of Technical Writing*, New York: St. Martin's Press, 1976.

3. *Chicago Manual of Style*, 13th Edn. Chicago: The University of Chicago Press, 1982.

4. *Concise Oxford Dictionary*, 6th Edn. Oxford, U.K.: Clarendon Press, 1976.

5. *Dictionary of Scientific and Technical Terms*, 4th Edition. New York: McGraw-Hill, 1989.

6. *Dictionary of Engineering*, New York: McGraw-Hill, 1984.

7. *Dictionary for Scientific Writers and Editors*. Oxford, U.K.: Oxford University Press, New York, 1991.

8. Heys, H. *The Preparation and Production of Technical Handbooks*, London, U.K.: Sir Isaac Pitman and Sons, 1965.

9. Horowitz, Joseph. *Critical Path Scheduling*, New York: The Ronald Press Company, 1967.

10. *IEEE Standard Dictionary of Electrical and Electronic Terms*, 3rd Edn. New York, NY, IEEE.

11. Jordan, Stello, ed. *Handbook of Technical Writing Practices*, New York: John Wiley & Sons, 1971.

12. Lee, Marshall. *Bookmaking: The Illustrated Guide to Design, Production, Editing*, 2nd Edn. New York: R. R. Bowker, 1979.

13. *Metric Practice Guide*, Canadian Standards Association, 1973.

14. Pathe, L. W. *Style Manual For Technical Writers*, General Electric Company, 1964.

15. Schoff, Gretchen H., and Patricia A. Robinson. *Writing and Designing Operator Manuals*, CA: Lifetime Learning Publications.

16. Shaw, H. *Handbook of English*, 3rd Edn. Toronto, Canada: McGraw-Hill Ryerson Limited, 1979.

17. *The Fundamentals of the Printing and Duplication Processes*, Domtar Newsprint Limited, Canada, 1973.

18. Turner, R. P. *Technical Writer's and Editor's Stylebook*, 1st Edn. Indianapolis, IN: Howard W. Sams and Co., 1964.

19. U.S. Government Printing Office. *Style Manual*, Washington DC: Government Printing Office, 1984.

20. *Webster's New World Dictionary of The American Language*, New York, NY: Simon & Shuster Inc., latest edition.

21. Weiner, E.S.C. and J. M. Hawkins. *The Oxford Guide to the English Language*. Oxford, U.K.: Oxford University Press, Oxford, 1990.

INDEX

A

Abbreviations
  correct use of 2
  for units of measure 236
  of/for
    titles of respect 204
    academic degrees 205
    words with symbols 218
    proper nouns, abbreviation of 202
  rules for 235–236
  using style guide for 180
Academic degree 205
Accordion fold 165
Acronyms 85, 206
Activity (PERT)
  number 26
  orientated plan 26
Addendum 132, 149, 150
Additional material 77
Adhesives, for binding 173
Adjustment
  and alignment 91, 140, 142
  points 59, 62
  procedures 25, 142
Administration, capitalizing words for 200
Adobe Illustrator 32
Advertising
  matter 77
  medium 48
Agencies 7, 193
Aircraft
  equipment 15, 101
  industry 158
Airmail papers 162
Aldus Pagemaker 32
Alignment
  preoperational 55
  guides (crop marks) on pages 158
  of paragraph numbering 82
  of text on paste-up 157
  procedures 4, 90, 140–142
Alternative wording 85
Ambiguity 115, 118

Amendment
  certificate 156
  control 156
  identification 155
  instruction 156
  issue 155
  level 155
  recording 105
  set 156
Ammonia-vapor 78–79
Announcement sheets 162
Annual reports 33, 110
Anonymous author, citing 240
Apostrophe, use of 245–246
Appendix
  accessories located within 132
  definition of 20
  for special procedures 132
  illustrations within 101
  in manual specification 73, 77, 93
  in PERT charts 29
  installation procedures in 17, 25
  listing in table of contents 105
  preparation details for 149–151
  sample page of 152
Approximations, in numbers 228
Arabic numerals 83, 121
Armed services, manuals for 147
Arrangement of technical manuals 10
Arrows 26
Artificial respiration 88, 95–96
Artwork, preparation of 107, 188, 191, 192
Assembly
  of material 26
  process 29
  views 107
Asterisk
  for footnotes 122, 242–243
  general use of 246
Atomic (proton) number 215
Attaching parts 91, 92
Attachment (see Appendix) 149
Audience
  for manual 19, 114, 177

for maintenance and repair 131
technical background of 181
Author(s)
cited in footnotes 239–240
technical, categories of 178

B

Backbone, of book, stapling 168
Bar charts 114, 117
Basis weight 164
Baskerville typeface 35
Bibliography 181, 277
Binder(s)
dividers for 29, 174
folded pages in 92
front covers, information on 97, 156
loose-leaf 168
special, for installation 17
specification requirements for 72, 78, 80
suppliers of 194
three-ring 87, 168
title pages for 88
typefaces, for covers of 180
typical, illustrations of 169
Binding, methods of 167, 168, 170
BITE 14, 62, 65, 137
Black bar, for amendment location 155
Blank pages 46, 83
Block diagrams 24–25, 90, 134
Body
of tab divider, text on 174
of table 122
of text, caution notices within 47
of type, legibility of 33
references in 86
Boldface 44, 79, 101, 207
Bond papers 162, 164
Book paper 163
Bookbinding 171–172
Booklets 162, 165, 168
Borders, removal of 158
Boxboard 163
Boxhead, for tables 122
Brackets (see also Parentheses)
213–214, 247
Bristol board 80
Brochures 48, 162, 165, 167–168, 171
Built-in test equipment (BITE) 14, 62, 137
Built-up fraction 211, 213
Bulk of paper 167
Bulk stock 7, 91, 143
Business forms 162

C

CAD/CAM 189, 191–192
Calibration checks 16
Camera-ready
copy, in specification 76, 81
copy, preparation of 2, 31, 157–159
documents 29
Capitalization, rules for 197
Captions 115, 118, 124, 159
Caption numbering 233
Cartoons, as illustrations 110
Case binding 167, 173
Cataloging of parts 143
Catalogs 168
Central documentation department 154
Centered emdash 122
Cerlox binding 87
Chapter(s)
breakdown of 20
definition of 20
examples of 40–41
first page of 42
headings for 42
layout of 39
numbering of paragraphs in 44
specification requirements for 78–83, 89
Chart(s)
bar type 114, 117
captions on 115, 233
constructing 118, 124
CPM
example of 30
flow 28–29
fault diagnosis 65–67, 91, 135, 138
illustrations as 107–114
manual structure 21
metric usage 226
pictorial
example of 118
type 114
pie type 110, 117
preparation of 2
specifications for 86
troubleshooting 22, 57, 67–69, 91
troubleshooting chart, example of 68
Chemical
abbreviations 235
elements, writing of 216
engineering, lexicons for 183
reactions 216
symbols 209
City of publication 240, 249
Clean (indoor) environments 80

Clear acetate, for manual covers 171
Client relationships 194
Clip-art 32
Codes, manufacturers, in IPB 91–92, 144
Collation 29
Colon, use of 44, 248–249
Color(s) of
  cover material 171
  index tabs 174
  printers 191
  photographs, control of 193
Column widths 34
Comma, uses of 115, 204, 231, 249
Commercial
  components 91
  hardware 16
  part 75
  photographic and reproduction 2
  purchasers 12
Common abbreviations without periods 236
Common nouns 202
Company
  address 88
  executive, as author 178
  name 48, 96
  policy 79, 196
  style manual, as guide book 180
  wordmark 88
Comparisons between figures 121
Complete equipment 60, 70
Complex table form 123
Component location 91
Component(s)
  commercial, in parts list 91
  definition of 75
  disassembly instruction of 86
  location of 91
  parts of a table 121, 144
  reference designation of 144
  specific references in figures 86
Compound
  adjective 256
  modifier 258
  noun 202, 258
  number 206, 228, 257
  verbs 256
Comprehension 61, 153
Comprehensive manual 9
Computer
  manufacturer 20
  repair facility 22
  software 110
Computerized parts control 143
Conclusions 107

Conflict 72, 76, 130
Conformance to policy, by editor 177
Conjunctions, in headings 83
Consistency
  in terminology 84
  organization 84
Constructive editing 177–178, 181–183
Consumable
  items
    in parts list 138
    servicing requirements for 138
  material, references to 86, 92
  part, definition of 75
Content(s)
  indexes to 46
  of front matter 46
  specified 74–82
  table of 24, 29, 88–89, 95, 101
    examples of 103–104
    in specifications 72–73
    list of figures in 124
  technical, of manual 2, 9–10, 180
  title of 84
Contract
  negotiations, in specification 76
  price of manuals 8
  staff 195
Controls
  and adjustments points 64
  descriptions of 11–12, 50, 55–57, 90,
    128–130
  summary of 12
Copy freeze date 72, 77
Copyright
  owner 77
  statement 155
Corporate identification, on manuals 180
Cost of production 179
Courier typeface 34
Cover(s)
  acetate material for 171
  book 171, 173
  layouts of 180
  marking of 78, 80
  page 95–97
  page, example of 97
  reference number on 156
  sheet, design of 29, 48
  stock, selection of 162
CPM flowchart 26, 30, 93
Critical path monitoring 26
Crop marks 158
Cross-references 86, 90, 144
Custom manuals 8–10

Customer specifications, for manuals 180
Cut-and-paste 110, 157, 159
Cutaway perspectives 190
Cutaway drawing 108, 114
Cutting
  marks (crop marks) 158
  out unneeded material 159
Cycles of the review 29

D
D-ring binders 168
Dangerous material 139
Dash, uses of 253
Data sheets, in appendixes 150
Day-to-day
  maintenance 7
  operational procedures 149
Decimal
  designations 219–220
  fractions 119
  multiples 119
  place, in numbering 44
  point 82, 219, 231–232, 261
Deféctive parts, location of 138
Definitions 19, 72, 75, 85
Denominator, in mathematics 211, 213
Depot repair 14
Derived unit 221, 226
Description(s)
  column, in parts listing 91–92, 143
  general 54, 90
  in table preparation 122
  of equipment 11, 50–57, 61, 90
  value of, for sales 61
Design changes 101, 195
Design philosophy, basic 55
Desktop publishing 19, 29–30, 122, 157, 189
Detailed manual structure 24–27
Development phase, of manual 76
Diacritics (accents) 207
Diagnosis tree 67
Diagram(s)
  block
    production planning 26
    functional 51, 61–62, 90, 135
    simple, of system 63
    as frontispiece 51
    simple 135
    theory in 20–21, 24–26
  conversion tables on 87
  CPM or network 26
  for technical descriptions 90–91, 142
  illustrated parts breakdown 146

large size, handling 165
schematic 55, 61–62, 90, 108, 135
Dictionaries, using 84, 175
Direct quotations 198, 251, 262
Direct references 84
Direct-image process 81
Discrete quantities 114, 115
Disassembly
  detailing method of 86, 107
  exploded view of 108
  instructions 69
  procedure, example of 139
  procedure, writing of 13, 60, 70, 140
Display form 107
Disposal of the old pages 154
Dividers
  as specified 80, 87
  indexed with tab 87
  tabbed 29, 78, 150, 174, 194
Divisioning of
  manuals into chapter/sections 19, 87, 89
  paragraphs 39, 44
  scales for graphs 119
  table of contents 51
Document(s)
  applicable, list of 74
  camera-ready 29
  conflict between 76
  with legal aspects 42, 71, 101
Domestic 22, 153
Dots, multiplier 220
Double dimensioning 87
Double-meaning abbreviation, avoiding 236
Draft
  editing of 181, 175, 183
  manuscript 1, 183
Draft manual
  numbering of 183
  PERT or CPM chart for 93
  specification requirements for 72, 76–78
  typing 188
Durability, of printing papers 162

E
Earliest date 26
Edited and formatted text 29
Editing
  care in 39, 177
  definition of 175
  level of 4
  marks, table of 182
  of technical manuals 4, 184
  procedures in 181, 183

Editor(s)
  characteristics of a typical 176–185
  marks used by 182
  role of 175–176
Effective
  date of issue 96
  pages, list of 88, 95, 105
Eight-page booklet 165
Electric shock, action and rescue 88, 95, 97
Electrical
  equipment 20
  hookups 16
  requirements 24
Ellipsis 254
Embossing 173
Emdash (see also Dash) 253–254
Emergency repairs 12, 13
Endash (see also Dash) 253–254
Engineering
  departments, as information sources 192
  drawings, using in manuals 158
  writers 178
Equation
  displayed 209
  numbering of 210
Equipment
  configurations 11, 132
  manufacturer 14, 16, 50, 77
  operation 55
  readiness 130
  supplied 24, 90
Errors and inconsistencies 184, 190
Estimates of time and cost 195
Ethnic groups 200
Evidence in litigation 133
Exclamation mark, uses of 255
Existing engineering drawings, use of 158
Exploded
  order of disassembly 70
  views 91, 108
  view illustration 16
Extensive realignment procedures 139
External
  printer 31
  reproduction company, using 159
Extraneous material, removing 159

F

Fabricated
  drawings 135
  parts 91
Factions, capitalization of 200
Factory-training, for customers 193
Fadeout 108

Failure of a component 57
False economy 196
Fan folded paper 165
Fault
  diagnosis 62, 91
  diagnosis chart 65, 67, 135
  location table 138
Faulty component 137
Field service 13
Figure
  and table callouts 183
  numbers 25
  omission of first, in dates 246
  separation of, in large numbers 252
  symbols 207
  titles 51, 73, 86
File cards 163
Filing and recording systems 190
Final
  assembly 92
  draft 29
  inspection 176
  manual 72, 76, 80
Fine paper 162
First aid
  and artificial respiration 88, 95–96, 98
  and rescue 24
  information 46
First word
  in a sentence, capitalization of 198–199
  in salutation of letter 203
  of a sentence, when a number 206, 230
Flowchart 26, 27, 65, 74
Flyleaves 173
Foldout
  drawings, numbering of 84
  page 45
Folding 164-166
Foldout drawing 45
Folio 45
Font 34, 42
Fonts, used in technical material 207
Footnotes 122, 180, 183, 246
Footnotes, general procedure for 238
Format
  common 49
  in specification of manual 74, 78, 80, 82
  of appendix 151
  of pages 39
  preparation of 2
Formality, of capitalized words 197
Formatted text 29
Four-page folder 165

Fractions
  8, 110, 115, 119
  expressed numerically 232
  of units of measure 115
  representation of 211–213
  using solidus 264
  written out 232
Front cover sheets 29
Front matter
  in a maintenance manual 61
  in an operation manual 50
  in specification 73, 83, 88–89
  introductory material in 95
  material in 24
  numbering of 46
  options listed in 150
  preparation of 50
  reference data in 49
  safety instructions in 99
  warranty in 51
Frontispiece 50–51
Full capitals, when used 203, 206
Fully-justified 39, 81
Functional
  block diagram 24, 55, 61–62, 90, 135
  controls, explained 61
  diagrams, example of 63–64
  diagrams, forms of 108, 110
Function of a product 214
Functions, mathematical, setting 213
Funds, limiting scope of manual 59
Furniture 191

G

General information 10–12, 24, 73, 89–90
Generalized figure 114
Generic terms, capitalization of 202, 205
Geographical areas, capitalization of 206
Glossary 85
Gluing 171
Government 7, 198, 200
Graduate technologist 5
Graduated theory 63, 134
Grain, of printing paper 81, 162–168
Grammatical mood 73, 84
Graph style illustrations 108
Graphic(s)
  consistency in manual sets 49
  method of illustrating 107, 114
  on camera-ready copy 127
  scanning drawings for 158

software packages 32
symbols, used in manuals 54
files 158
presentation of equipment 121
representation 51
symbols, specified 74
Graphs 107, 115, 119, 121
Greek symbols 207, 217
Guarantee (warranty) page 88
Guide lines on pasteup card 157
Gutter (margin) 39, 167, 171

H

Halftone reproductions, paper for 163
Hand-drawn characters 108
Hand-inserted changes 154
Handbooks
  contractual arrangements for 193
  jurisdiction for 188, 193
  manufacturer's 22
  military 46
  technical
    production of 188
    writing 181
  time and cost estimates for 195
Handwritten
  amendments 78
  lettering, identification of 208
Hard drive 8
Hardbound covers 172
Hazardous
  material 24
  situations, amendments to cover 153
Headings
  capitalization of 199
  for safety notices 85
  general, in specification 73, 79–83
  in tables 122
  numbering of 44
  style of 33, 39–40
  typefaces used for 180
Heavy-duty staplers 168
Helvetica 34–35, 81, 88, 121
High quality paper, for illustrations 158
Hole reinforcements 80
Hours-of-operation schedules 138
Humorous characters 108
Hyphen (see also Dash) 253, 256
Hyphenation
  of written numbers 228
  readability of words with 39, 42

I

Ibid., use of 236, 241
IBM-clone, computer type 8
Identification
  by graphics 107
  data presented 50
  date, of manual 51
  ease of, tabs for 87
  of
    amendments 155
    appendixes 151
    covers, by embossing 168, 180
    figures, by title 86
    parts 70, 108
Identifying characteristics of parts 144
Illustrated
  assembly parts list 143
  parts breakdown 25, 70, 143, 146
Illustration(s)
  as parts breakdown 25, 70, 143
  combining with text 19
  disassembly of equipment, example 139, 141
  requirements and definitions for 107–119
  in
    appendixes 101
    front matter, listing of 48–49, 101
    maintenance sections 142
    operation sections 24
  of equipment 16, 24
  of test setups 25
  paste-up for 157
  positioning of, in manuals 33
  reference designations in 144
  reworked, example and procedures for
    158–159
  specification requirements for 78–81, 86–91
  using and preparing 55–58, 158
  using CAD/CAM for 192
  of the system 24
Illustrator 32, 107, 110, 118, 190, 193
In-house manual 31
In-plant training 6
Inadequate warnings, hazards of 133
Incandescent lamps 75
Incomplete data for manual 59
Inconsistencies 179, 184, 192
Incorporation of revisions 155
Incorrect
  maintenance, causing malfunctions 153
  reassembly, problems of 139
Increments, in charts 114
Index
  dividers 174

in table of contents 46, 49, 89
numerical, in parts lists 143–144
  example of 147
reference designation 25, 143–144
  example of 148
Indicators, summary of 12, 130
Individual components 10
Individualized 8
Industrial 4, 5, 8, 14, 16, 84, 179, 193
Informal tables 123–124
Information
  artificial respiration 96
  first aid 44
  for amending manuals 153–155
  for PERT charting 26
  for servicing manuals 22
  from CAD/CAM sources 192
  general system, in manuals 10–12, 24,
    89–90, 132–133
  in appendixes 149–150
  in specification appendixes 93
  modes of operation 59, 90
  on tables 51, 121–123
  on charts 114–115
  on divider tabs 174
  on illustrations 107
  on manual covers 168–169
  sales, in manuals 54
  spares 50
  technical, for operators 23–24
  trouble analysis 25
  troubleshooting 58, 61, 67, 90
  warranty 24, 46
Inspection schedules and procedures 137
Installation
  data sheets for 150
  illustrations for 107, 149
  instructions section 10, 16–22, 25, 50
  material from CAD/CAM sources 192
  requirements in specification 73, 79, 89, 92
Instructions
  for incorporating amendments 154
  for parts removal 139
  for unwrapping equipment 16
  manuals 7–8
  safety precautions, and 96, 133–134
Insulation material, in parts lists 143
Intended audience 177, 183
Interface boards 8
Interim manual 72, 76, 79
Interleaf (software) 31
Internal
  control of manuals 154
  reports 187

International System of Units (SI) 217
Introduction, as basic element of manual 50, 54, 61
Introductory
  material 95
  statement 9, 198
IPB (illustrated parts breakdown) 25, 70
ISO A4 paper (metric) 39
Isobaric nuclides 215
Isobars 215
Isotopes 215
Isotopic nuclides 215
Issue number, of manual 96
Italic 35
Italics, in technical manuals 207–209
Itemized checklist 16

J
Job cards, to control work 195
Journalism, background for writer 5, 176
Journals
  medical and chemical 183
  citing in footnotes 240
Justified text 38

K
Kerning, examples of 36–37

L
Labor and money 9
Languages 128
Large number(s) 7, 9, 49, 230–231
Laser printer 1, 81
Latest start date 26
Latin, usage of 207, 236
Lawyer (see Legal)
Layout, of
  manual 24, 32–33
  pages in a manual 38
  pages, using computer template 157
  parts lists, in specification 91
  publishing department 190
  table of contents 101
  text, in specification 84
  title page, in specification 88
Leaders (dotted lines) 124
Leading (type)
  explanation of 36
  illustrated 35
Ledger paper 80
Left-hand (verso) page 47, 157

Legal
  counsel, warranty aspects 51, 101, 133
  elements of specifications 42, 71
Legibility 33
Length of text lines 34, 37–38
Letter(s)
  distinguishing handwritten 208–209
  in symbols, spacing of 219
  omission of, in offensive words 247
  salutations in 203
Letterhead paper, in manuals 48
Lexicons 180, 183
Liability of a company 133
Library-quality hard covers 171
Limitations of
  equipment, detailed in manual 90, 107
  manual development 58
  of typesetting characters 207
  software packages 33
  space, for illustrations 119
Line
  drawings 12, 55, 107–108, 110, 116
  fraction 211–213
  horizontal, in derived units 221
  in footnotes 238–239
Linen 163
List of
  effective pages 88, 95, 105
  figures 101
  illustrations 50, 88, 95
  tables 50, 88, 95, 105, 121, 124
Litigation, evidence in 133
Loc. cit. 242
Localized repair 22
Location of parts, illustrations for 108
Logistics support 75
Logotype 48, 88
Loose-leaf (mechanical) binders 168–169
Lost production 12
Lowest level of service 60, 62, 69, 135
Lubrication points 91, 108, 142

M
Magazines, binding of 167–168
Mailers, paper for 163
Main
  body of text, in specification 86
  paragraph headings 44, 73, 83
  textbook 23
  frame computer 8, 22

Maintenance
  activity 65
  aspects 55, 78
  information 9
  instructions 10–13, 90, 131, 133
  manuals 13, 19, 55, 58, 61
Major
  equipment drawings 92
  overhaul 8
  parts of manuals 10–11, 14, 49
Manual(s)
  amending of 105
  commercial, spares listed in 70
  covers 10, 88, 162
  for US government 147
  planning 11
  planning stage 11
  printing of 161
  production plan 26
  set 11, 22, 156
  specification, sample 71
  structure 20–21, 25, 73, 89, 158
  structure chart 25
Manufactured part 76
Manufacturer's
  codes 92, 144
  designation of items 135
  handbooks 22
Manuscript 1–2, 4, 180–181
Maps, as illustrations 107, 129
Margin (gutter), for binding 171
Margins, removal of 158
Mass
  (nucleon) number 215
  number 215–216
  units of 226
Master manual 17
Mathematical
  designations and symbols 217
  expression 208, 255, 260
  operators 207
  signs of relation 207
  symbols 208
Matt surface 162
Mean time between failure (MTBF) 15
Measurements 36, 73, 87, 123
Mechanical
  components, inspection of 137
  drawings 25, 73, 89–92
  engineering 5
Medical journals, styles of 183
Menus, papers for 162
Microsoft Word 33, 157
MIL (military) specification 75

Military
  handbooks 46
  specifications, in IPB preparation 147
Minor
  components 55
  servicing 13, 50, 60
Misalignment of punched holes 174
Mode, standby, noise of 191
Model
  designation 8
  serial numbers 143
Modes of
  operation 9–10, 57, 134
  the equipment's operation 134
Modification leaflets 187
Moisture, in printing paper 163
Monitoring 11, 26, 91, 139
Most suitable layout 25
MTBF 15
Multiple of an SI unit 222
Multiplant companies 15
Multiplication sign (dot) 220
Mylar
  facing/tab reinforcing 174
  laminated tabs 80

N

Name(s)
  of month 236
  of seasons, expressing 202
Narrative text 78
Narrow column 39, 42
Negative
  exponents, representing 211
  powers, representing 221
Network diagram 26
Neutrons in the nucleus 216
Newspaper, titles of 175, 201
Next higher assembly 144
Node numbering 26
Nodes 26, 29
Nomenclature 73, 85, 123
Non serif font 121
Non
  -operational controls 57
  -standardization of manuals 77
  -technical (staff) 22, 24
Notation of amendment, in small print 155
Notices, warning/cautionary 47–48
Noun, capitalization of 197
Nuclear physics terminology 216
Nucleon number 215–216
Nuclides 215

Number of atoms in an entity 216
Numbering,
  of drawings 46
  in dates 232
  machine screw threads 234
Numerator 211, 213, 221
Numerical
  index 143–144, 147
  order 25, 144

O

Oblique stroke, use of 46, 221–222, 264
Observations 65
OEM 20, 22
Office
  space 188, 190
  stationary 162
Offset printing 158, 163, 171
On-the-job-training 24
One-digit numbers, rules for 227
One-of-a-kind equipments 8
Op. cit. 241
Operating
  and performance specifications 24
  characteristics 24
  controls 50, 55, 57, 59, 90
  instructions 9–12, 22, 24
  limitations of equipment operation 90, 129
  procedures 8, 24, 90, 129, 138, 177
Operation
  manuals, basic elements of 50
  manuals, illustration for 107–109
  modes of 134
  procedures of 127
  requirements, in specification 90
  theory of 10, 12, 24, 50, 55, 61, 127, 134
Operator
  error 57
  functions, maintenance procedures as 127
  level functions, troubleshooting as 90
  mathematical 207
Opinions, words conveying 54
Options 2, 7–8, 11, 24, 150
Order
  of footnotes 239, 242
  of importance of paragraphs 183
Ordinal numbers, rules for 229
Original equipment manufacturers (OEM) 20
Oscilloscope waveforms 108
Outside agency, using 189

Overhaul
  as a separate section 89
  function 16
  in maintenance manual 11–12
  periodic 13
  repair and, as a section 92
Owner's manual 22, 128–129

P

Page
  layout 32, 34, 38–39, 42, 88
  numbering 31, 33, 45–46
    in appendixes 151
    of figure listings 101
    of table of contents 101
    of table listings 121, 124
  replacement 54
PageMaker 32, 157
Pagination 45–47
Paper
  folds 166
  types (grades) 1, 48, 162
Paragraph
  headings 83
  numbering 42, 44, 47, 73, 82, 101, 151
  spacing 39, 73, 83
Parallel folding 164–165
Parchment, paper 162
Parentheses (see also Brackets)
  applications of 247
  for codes and part numbers in IPBs 143
  for next higher assembly 144
  in mathematics 213
  in parts lists 144
  in tables 115
  oversize, in mathematics 213
Part
  and section headings 73, 82
  in numerical list of parts 144
  numbers
    in parts lists 25, 143
    of manufacturer's equipment 91, 143
    of Mil–spec components 143
Partial schematics 86
Pasting, machine 165
Patents 191, 193
PCB cards, descriptions of 134
Pen-and-ink drawing 110
Percentage designations 233
Perforating, automatic, by machine 165

Performance
  checking 142
  criteria 10
  figures 150
  monitoring and checks 91
  specifications 24
Period, the 218, 220, 260
Periodic
  and preventative maintenance 138
  overhaul 15
  service and inspection 13
  servicing 9
Periodical, identifying by reference 262
Personal
  computer 8, 20, 32, 157, 171
  titles 204, 236
PERT flow charts 26, 93
Phonetic symbols 207
Photo
  copying 158
  mechanical transfer (PMT) 159
  offset 81
  copied reproduction 2
Photographic paper, illustrations on 159
Physical characteristics 11
Physical dimensions 24
Pica, defined 35
Pictorial charts 114
Pie charts 110
Piping drawings 92, 108
Plastic (Cerlox) binding 170–171
Plasticized card 96
PMT (photo-mechanical transfer) 159
Point(s)
  of a compass 206
  size of typefaces 34–36, 81, 88
Point-to-point wiring 108
Political groupings 201
Political parties 200
Post-overhaul testing 92
Power
  electrical, in installation manual 92
  requirements, in manual 11, 55
Preinstallation preparations 92
Pre-operational adjustments 55
Preface 46, 50–51
Preliminary knowledge 135
Preparation (of)
  amendments 154
  appendixes 151
  camera–ready 48
  detailed manual structure 24–27

errors in initial (draft) manual 105
final manual, for specification 80
for binding books 172
format 2
manuals for US government 147
parts listings 70
plan for manual production 19, 26, 29
proposal and brochures 4
publication plan for technical manuals 93
publications, experience in 31
systems, various, for typesetting 31
tables 121–125
technical manuals 4, 71
text 38, 47–48
text pages 157
Prepositions, in headings 83
Presentations 162, 180
Preventative maintenance 12, 15, 25, 90,
  138, 176
Principles
  of editing 181
  of operation 58, 90, 134
Printing
  copy freeze date, prior to 77
  corrections, prior to, in specification 76
  costs 9
  letterheads and logotypes 48
  of manual 161
  of text, in specification 81
  paper, selection of 161
  process 29, 48, 73, 81, 162
  process, in CPM charts 29
  process, for manuals 162
  proportional spacing in 36
  surface, requirements for 162
Procedures, troubleshooting 65
Product (multiplier) signs 208
Production
  plan 26
  sequence 185
Professional
  engineering writers 178
  title 204
  quality binding, on small scale 171
Profit 12, 15, 176
Profit-and-loss 12
Projected pagination 47
Promotional material 107, 162, 165
Proofing 179
Proper noun 201–203
Proportional spacing 36
Proposal and brochure preparation 4
Proton number 215–216

Publication(s)
  cited, in footnotes 240
  plan 73, 77, 82, 93
Publicity departments 191, 193
Publicity printing 162
Published document, value of 175
Punching, for binding 170, 174
Punctuation, of text 2–3, 181, 184
Punctuation, rules of 245

Q

Quality
  control 176
  of writing 33
  of paper for laser printing 81, 158
  of technical manual 1, 9
Quantitative relationships, in pie charts 110
Quantity (of parts) used 144
Quark XPress 32, 157
Question mark, rules for 259, 261, 263
Quotation marks 240, 253, 255, 262
Quoted material, wording of 198, 262

R

Rag fibers 162
Radicals, rules for 214
Ragged right 39
Rational fraction, form of 212
Ratios 234
Readability 3, 34–38, 149, 162
Realignment 140
Ream 164
Reassembly 60, 69, 107, 139
Recommendations 22
Record sheet 89, 142
Recto (right-hand) page 42, 47
Reduced illustration, text on 158
Reference
  designation index 25, 143–144, 148
  numbering 238
  work 161
Regional terms 206
Regular (periodic) inspections 137–138
Reinforcing 172, 174
Related statistics, in tables 121
Reliability 11, 15, 49, 189
Remedial action 57, 67, 196
Removal and replacement 25, 91, 139
Repair
  and maintenance section 131–142

and overhaul 12–13, 59, 73, 89, 92
and replacement 92
without reconstruction 59
Repairable subassembly 135
Replacement
  of assemblies 91
  pages 89
  parts, list of 70
  procedures 13, 25
Report writing 4–5
Reprints 162
Requirements
  measurements, in specification 87
  electrical, of equipment 24
  in general description of equipment 55
  manuals, for foreign countries 75
  of typefaces and selection of 33–34
  tariff and import 61
Rescue 24, 88, 95–96
Residual voltages, hazards of 139
Restrictions 2
Revision(s)
  block 105, 154–155
  during writing 29
  to manual 9
Rework 158–159
Right-angle fold 165
Right-hand (recto) page 151, 157
Roman typeface 34–37, 81–87
Roman font for math functions 207–209, 213
Round numbers, rules for 228
Routine maintenance operations 135
Rules of grammar 3

S

Safety
  activities, using cartoons 108
  features, detailed in manual 128
  instruction page 99
  instructions 96, 99, 133
  measures, legal aspects of 133
  officer 133
Sales
  brochures 48
  organization 7
Salutation, in letters 203, 249, 252
Sans serif, font type 35
Sans serif, in scientific writing 207
Scales for graphs, choosing 119
Scanned copy 160
Scanning device 158

Schedule 26, 29, 137, 181
Schematic diagrams 55, 74, 90, 135, 137
Schoolbook typeface 35
Scientific
  names of genus 201
  papers, preparation of 216
  terminology, presentation of 2, 207
Scope of manual 58
Scoring, for bending 165
Screened copy, paper for 163
Screw threads, designation of 234
Second person imperative mood 84
Secretary, technical 188
Section(s)
  definition of 20
  dividers for in a specification 78, 80
  first page of 42
  in multivolume manual specification 87
  installation 16
  maintenance and repair 131–142
  of a manual 33
  operation section 127–130
  page numbering in 45, 82
  paragraph headings in 83
  parts list (IPB) preparation 143–148
  requirements (specification) 89–92
Sectioning of manuals 49
Selection of manual type 61, 161
Selection of typeface 34
Semicolon, use of 263, 264
Seminars 6
Sensitive material 46
Separate overhaul manuals 16
Sequence of pages 172
Serial numbers 50, 143, 233
Serif, in fonts 35, 121, 207
Service
  department 12, 57
  field 13
  field-shop 14
  guarantee (warranty) 52
  lowest level of 137
  periodic inspection, and 65
  personnel 131
Sewn binding 167, 172–173
Shadowless print 159
Sharp detail, in illustrations 108
SI
  base units 223, 226
  derived units with special names 223
  measurement system, rules for 217
  prefixes, list of 224

system, other units used in 225
Side-stitching 167–168
Signs of relation 207
Signatures 165–168, 172–173
Six-page folder 165
Size
  and weight descriptions in manual 55
  of cut paper 164
  of illustrations 158–159
  of job, for collating 185
  of paper and binders, for specification 80
  of typefaces 33–36, 42
  wordmarks and logos, in specification 88
Slang 201
Slitting of folded pages 165
Software
  manuals for 22
  programs, useful for manual preparation
    Adobe Illustrator 32
    Aldus Pagemaker 32
    Microsoft Word 32, 157
    Quark Xpress 32
    Ventura Publisher 32
    WordPerfect 32, 157
    Zerox Docubuild 32
Soilage 161
Solid drawings 108
Solidus, use of 211–212, 221, 245, 264
Source line 122
Spare parts 16, 70
Special
  design, part made to 76
  instructions, in maintenance 10, 140
  parts, handling of 76
  tools 13, 54
Specialist 3
Specially written manual 8
Specialty equipment, manuals for 8
Specifications
  as general information 11
  chemical 55
  customer's 180
  electrical 55
  index to 72
  mechanical 55
  military, for IPB preparation 147
  of consumable products, need for 86
  preparation of, for manuals 71
  requirements, for manuals 78–93
  sheet 8, 26, 54
  summary of technical 22, 26, 54
Spelling 3, 84, 177
Spine strip, for cover 29

Spiral and plastic binding 170–171
Standard
  catalog item 75
  commercial hardware 16
  manual 9
  part 75
  printing practices 83
Staples 167–168
State-of-the-art 11, 75
Stitching methods 167–168, 171
Stop-gap publication 187
Strength of paper 162, 173
Structure
  chart 25, 148
  chart, of typical manual 27
  of basic manual, block diagram of 21
  of manual, determination of 20, 89
  of page 33
  sentence, as an informal table 123
Stub 122–124
Style manual, value as reference 179–180
Styles 34–35, 167, 171
Styles of reference numbering 238
Subassembly
  defined, in specification 75
  faulty, determination of 57, 135, 138
  functional diagrams for 61–62
  lack of technical data for 22
  listed, as spare 12
  operation, explained 24, 61, 135
  parts list, requirements for 91–92
  quantity of parts for 143–144, 150
  references to, in drawings 150
  repair of 69
Subdivision(s) of
  levels in table of contents 49
  headings 11, 39, 44, 73, 82–83
  manual, explained 19–20, 75, 89
  paragraphs, by numbering 44
Subheadings 11
Subparagraph(s)
  defined 20, 42–43, 73, 82–83, 101
  heading form for 42
  numbering of 44–45, 82–83
  typefaces for 42
Subscript(s) in
  metrication 217
  mathematic typesetting 209, 212
  scientific terminology 215–216
Substitution of amended pages 153
Sulfite fibers 162

Summary 12, 22, 24, 130, 199
Summation, using parentheses for 213
Sums of money, expressing 231, 261
Superscript, in
  tables, for footnotes 122
  mathematics 209, 212
  scientific text, typesetting 215–216
  footnotes 238–239
  metrication, rules for 217
Supplementary
  information in footnotes 237
  networks 26
  titles 81
Symbols
  graphic, special, in manual 54
  in illustrations, identification of 119
  in tables 122
  reference, creation of 192
System
  block diagram 61, 90
  information 24
  instruction manuals 11–12
  theory descriptions 57

T
Tabbed dividers 29, 78
Table numbering method 46
Table of contents, layout of 101, 103–104
Tables, preparation of 121
Tab divider sheets 80, 87, 158, 174
Tabulated matter 51
Technical
  author(s) 177–179
  documentation 32, 185
  illustrator 107, 190
  literature, as basis of manual 187, 189–190
  manual(s)
    arrangements of 10, 39
    binder, types for 168, 171
    diagrams, large, in 159
    editing and preparation 4
    illustrations for 107
    photographs in 193
    specification, preparation of 71
    structure of 20
    tabbed dividers for 174
    text, layout of 39
    titles of 48
    typography for 33–34
  material 1, 3, 25, 33, 76
  specification sheet 24

staff 23–24
summary 22
writer 2–6, 96, 187–190
Technician training 23
Technician-level activities 57
Technologist 5
Temperature
  in SI units 219, 223, 225
  levels, expressions of 234
Templates, for page layouts 157
Temporary staff 195
Terminology
  mathematical and scientific 2
  typographic 34
Test
  and alignment 25, 89–91
    bench 12, 62, 142
    equipment 13–14, 57, 70, 90, 135–137
  equipment data 25
  points 61
  setups 25
Text
  handling 32
  page preparation 157
  pages, numbering of 83
  writing of 26
Texture of paper 34
The cover page 96–97
Theater programs 168
Theory of operation
  detailed requirements of 12, 24, 60–61, 134
  drawings in 142
  manual requirements for 10, 22
  specification requirements for 73, 90
Thermal binding 171–172
Thickness of binding materials 80, 87,
  167–168
Third person indicative text 84
Three-ring binders 87
Time
  and cost factors 195
  designations 234
  intervals 15, 149
Times roman 34–41, 81
Timesheets 195
Title
  heading 158
  of works, captions and heads 199–202, 239
  page(s) 50–51, 83, 88
  sheets 80, 162
Tool shop 14
Tools required 92
Total quantities 144

Training
  class 76
  manual 131
  material 12, 23
  programs 79
Trimming 165
Trouble
  analysis 25, 91, 138
  diagnosis 67
Troubleshooting 16, 50, 57, 59–60,
  65–69, 138
  chart 68
Two-column page 39
Type
  leading, example of 35
  leading of 36–38
  of folds for booklets 165–166
  of paper for manuals 161–164
  size (see Fonts) 34–35
Typed material 208
Typeface
  for math functions 213
  in footnotes 239
  Latin and Greek 207
  roman, for subscripts 209
  roman versus italic 208
  styles 33
Typesetting characters, in math 207
Typical manual structure chart 21
Typographic
  relationships 34
  presentation of equations 207

U

Uncoated book paper 163
Undergraduate 5–6
Unessential details 158
Unit
  of measure 115, 122, 125, 209
  reference designation index 144
Univers typeface 35, 81, 88
Unpacking instructions 92
Unretouched originals, as illustrations 79
User-training, manuals for 23

V

Values in degrees (metric) 219
Vehicle owner manuals 129
Vellum paper 163
Ventura Publisher 32, 157

Verso (left hand) page 42, 47
Vinyl binder material 80, 173
Volume(s)
 definition of 19
 multi, in specification 73, 78, 87
 single, for instruction manuals 13
 table of contents for 89, 101

W
Warehouse and labor costs 9
Warning 85, 99, 128, 133–134
Warranty (garantee)
 information 24
 page 88, 95

statement 51, 101–102
Waveforms 108
Weight of
 of equipment, detailed 11, 24, 55
 of typefaces 34
 paper 81, 162–164, 171
Wide column page layout 39
Wire-stitched brochures 168
Wiring diagrams 16
Word-processing 2, 31, 188–189
WordPerfect 1, 32, 157
Worker safety 133
Workstation 31
WYSIWYG 31